向 2015 国际光年致敬

同济大学建筑学专业"建筑物理（光环境）"
教学成果专辑

2015·流光魅影
LET'S BE YOUR LIGHT

郝洛西 等 著

同济大学出版社
Tongji University Press

目录
Contents

Part 1	Prologue	写在前面	6
Part 2	Teaching highlights	教学花絮	18
Part 3	Task introduction	作业简介	22
Part 4	Rehearsal	彩排现场	36
Part 5	Final showcase	作品展示汇报	45
01	Overture	光轮	48
02	Think of me	漩光	56
03	Angle of music	冰影雾缭	62
04	Music of night	夜辉律动	70
05	All I ask of you	两生花	78
06	Wishing you were somehow here again	追忆似水年华	86
07	Wandering child	幻影移形	92
08	The point of no return	不归路	100
09	Finale	仲夏夜之星	108
10	Masquerade	化妆舞会	114
Part 6	Curtain calls and award	谢幕与颁奖	122
Part 7	Review of years of teaching	历年教学回顾	124

Light and shadow, heart and shape

Ever since the existence of light, numerous shapes of shadows follow.
Ever since the existence of architecture, light and shadows accompany it.
As the Chinese sayings go: "Hou Yi shooting the suns", "Li Bai staring at the moon" – natural light gives buildings vivid lives; "Ying people holding the candle", "Watching lanterns on the Lantern Festival" – artificial light expands the horizon of cities and architecture. In the industrial society, "Edisons" lit up the night sky and every corner of buildings with electrical lights; while in the information society, new technologies emerge. LEDs, with less energy consumption and more flexibility, emit even more beautiful light and colors, leaving people soul-stirring light and shadow.
To help future architects understand the charm of light and shadow, Professor HAO Luoxi and her team have been teaching the course "Architecture Physics: Lighting Environment" for 15 years. This is a fundamental, 17-hour course for sophomores of College of Architecture and Urban Planning (CAUP), Tongji University. The course hour is short, however, both teachers and students exhibited great involvements, and very much enjoyed the wonderful teaching and studying processes. The three books presented here in front of us today, including "2015·Let's be Your Light", "2014·Light Shadow and Season" and "2013·The Light of The Antarctic", are the achievements from cooperation between teachers and students during the years 2013 – 2015.
We are moved by such achievements. These are teaching and experiments with rich characteristics of the time. They are combinations of art and technology, through cooperation between teachers and students.
They think and work: arrange light sources, design devices; they learn from history and raise questions to future, wandering around in the free realm of light and shadow, being intoxicated. They organize exhibition visits, design dramas, prepare music, and turn pure lighting technology into comprehensive art exhibitions as well as lighting festivals for teachers and students.
All of these stemmed from their desire to pursuit technology improvements, and their wish mood to seek various art forms. The forms to express vision and longing comes from their desire and wish.
We look forward to seeing Professor HAO and her colleagues to continue the innovations and "emit more gorgeous light". We appreciate people who offered their help during the series of teaching processes. We remember the warm shadows they left, and wish students great achievements in future study and creation, and "exhibit more beautiful colors".

Prof. Dr. LI Zhenyu
February 22, 2016 (The Lantern Festival)

光与影，心与形

自从有了光，便有了千姿百态的影。

自从有了建筑，光影便一直与之相伴。

后羿射日，李白对月，自然光赋予建筑有声有色的生命；郢人秉烛，元宵看灯，人工光拓展了城市和建筑的时空。进入工业社会，爱迪生们用电灯照亮了夜空，照亮了建筑的每一个角落；进入信息社会，新技术层出不穷，LED 以更少的能源，更便捷的组合，发出更加美妙的光色，留下更加动人的丽影。

为了让未来的建筑师们理解光影的魅力，郝洛西教授与她的团队开设"建筑物理（光环境）"已经有十五年了。这是一个 17 课时、面向同济大学建筑系本科生二年级的专业基础课程。课时虽少，但参加课程的老师、同学们却非常投入，非常享受这样美好的教学过程。今天呈现在大家面前的三本书《2015·流光魅影》《2014·光影时节》《2013·南极之光》，就是 2013–2015 年期间师生共同合作的成果。

这样的成果，让我们为之感动。这是富有时代特征的教学和实验，是艺术和技术的结合，是老师和学生的合作。

他们动脑动手，布置光源，设计装置，向历史学习，向未来提问，徜徉在光和影的自由王国里，如痴如醉；他们组织展览，策划剧情，配制音乐，把单纯的光的技术，变成了综合的艺术展示，变成了师生共同的光影节日。

这一切，都是因为他们有追求技术进取的心愿，追求艺术多态的心境。有了心，才有了表达情怀的形。

我们期待郝老师和她的同事们不断创新，发出更加绚烂的光；我们感谢给予这一系列教学活动各种帮助的人们，记住了大家留下的热情的影；我们祝愿同学们在未来的学习和创作中大有作为，呈现出更加美丽的光与色。

院长
2016 年 2 月 22 日元宵节改定

Inheritance and development

The book series on your hands belongs to the series of Tongji "Architecture Physics: Light Environment", which is the outcome of fifteen years' teaching exploration and innovation. This book, for the first time, attempts to introduce the format of digital media besides the paper-based publication. Readers can download the APP and have an intuitive experience of the review site, and fully appreciate the splendor of each work in the books. This series includes three books: the first book is "2015·Let's be Your Light" published in 2015; the second book is "2014·Light Shadow and Season" in 2014; the third one is "2013·The Light of The Antarctic" in 2013. Due to the limited space, teaching achievements before 2012 are not included in this publication; however, years of work still exudes the unique light, such as the large-scale installations"sound and light show" during the 2010 Shanghai World Expo; the work "bloom", after the demonstration in Expo cultural center, continues its applications in healthcare buildings, passing the infinite charm of light and space.

As an important fundamental course, "Architecture Physics: Light Environment" is designed for the sophomores with an architecture major. Given only 17 credit hours, the course is taught in three main aspects: (1) light, color and visual environment; (2) the light source and lamps selection (laboratory teaching); (3) quantity and quality of lighting. As there are courses "Interior Lighting Art" and "Daylight and Architecture" for senior students, this course can be seen as the entry-level courses for students to have a feeling of light and illumination. Over the years, the coursework topics are carefully designed by the teaching team, focusing on the light and optical properties of materials, shape, light-emitting device, the control concept, light drawings and expression, and other rich and comprehensive understanding. This helps students to have innovative thinking, practical hands-on operation, and the attitude of pursuing excellence in design works. As a team, students cooperate to complete the job. Everything is designed for students' self-learning, exploration, and practice.

Although the course is limited to six weeks and three credit hours per week, the students are sophomores who just stepped into the design world, and furthermore, the students must ensure that their main courses are not much affected, the students succeeded to finish the good work in the form of team work, with the limitation of defined theme, venue, time, number and color of LEDs, the conditions of environmental protection waste materials. The last day of teaching is for the show and assessment of homework. Therefore it can be treated as a time-limited quick architecture design, all the on-site dynamic changes are students' "hand control". The huge number of constraints are indeed challenges to both teaching and learning sides, and this is why I, as the teacher for this course, always be amazed by students' works.

We appreciate the support from the president of the Tongji University Press, Prof. ZHI Wenjun. He always pay close attention to the team's teaching efforts. Prof. ZHI have suggested us to publish our teaching efforts many times. Due to the fact that it is really difficult to reflect the content and expression of students' coursework by paper-based publications, this effort had been delayed to seeking the opportunity of digital-media publication. This summer, Prof. ZHI and the president of Jin Shang Digital, Mr. Xiaobo Xing, had a discussion with me at the Tongji University Press regarding the publication of this teaching effort. Ultimately, we decided to utilize both paper-based and digital-based formats to publish the three books, to share our experience and results from our 15 years of teaching on the course "Architecture Physics: Light Environment" .

Special thanks to CAUP colleges, graduate and undergraduate students, and friends from industry who participated in this teaching work. Especial thanks to LIN Yi and CUI Zhe, for their contribution in curriculum construction.

As 2015 is the "International Year of Light" declared by the United Nations, to commemorate "the father of optics" - Ibn Hazm, who wrote an important optics book a thousand years ago, and a series of important finding in optics history, we would like to take the opportunity to pay tribute to the "International Year of Light"!

HAO Luoxi
12/15/2015

传承与开拓

您手中的这套书是同济大学"建筑物理（光环境）"这门课程15年教学探索与创新实践的成果汇编。这套书尝试在纸媒出版的基础上引入数媒，读者下载APP就可直观体验到作品评审时的现场情况，充分领略每个作品的光彩。本套书共计三册，第一册是2015年的作业汇编《2015·流光魅影》；第二册是2014年的作业汇编《2014·光影时节》；第三册是2013年的作业汇编《2013·南极之光》。由于出版篇幅所限，2012年之前的教学成果没有列入此次的出版计划，但那些年的作品依然散发着特有的光芒，如2010年上海世博会期间的作业"声光SHOW"大型装置；再如当年的作品"绽放"，历经世博文化中心的演绎应用，作为后世博的再应用，继续在医疗健康建筑中，传递着光与空间的无限魅力。

作为建筑学一门重要的专业基础课程，"建筑物理（光环境）"面对建筑学专业二年级的本科生授课，鉴于只有17个学时，主要讲授三个内容：光、颜色与视觉环境；光源与灯具选型（实验室授课）；照明的数量与质量。对四年级本科生开设有"室内照明艺术"和"日光与建筑"，因此可以看作是同学们接触光与照明知识的入门课程。历年课程作业选题来自教学团队的精心设计，重点在光、材质及光学特性、形态、发光器件、控制概念、光照图式及表达等丰富而综合的理解，培养学生开拓创新的思维能力、实践操作的动手能力、追求卓越的设计态度。作业完成以小组为单位，培养学生协同攻关的能力。一切向着自主学习型、设计探索型和实践应用型的学习方式迈进。

尽管只有为期六周、每周三个学时的授课，尽管他们只是初涉设计的二年级学生，但他们在保证专业课程设计作业不受影响的情况下，以团队合作的形式，在限定主题、场地、时间、LED颗粒数量和颜色、环保废弃材料的条件下，完成最后的作业。教学的最后一天进行作业展示及评审。因此也可以说，这是一个类似建筑设计快题的限时作业，现场所有的动态变化均是学生们的"徒手控制"。如此多的限定条件，对教与学来说都是莫大的挑战，我作为任课教师和大家一样，对学生们最后呈现的作品总是每每发出由衷赞叹。

承蒙时任同济大学出版社社长的支文军教授一直以来对团队教学的关注，多次讨论将该教学的成果汇编成册出版，但苦于仅凭纸质的确很难全面反映该课程学生作品的内容与表达，于是就心有不甘地一直寻求数媒出版的可能性。今年暑假期间，支教授连同"今尚数字"邢小波先生约我到同济大学出版社，一起探讨出版事宜。最终我们决定以纸媒加数媒的出版形式，由张睿老师担任责任编辑，出版这三册书，分享15年来承担的"建筑物理（光环境）"课程的教学成果。

在此特别致谢团队中共同进行课程建设的林怡、崔哲两位老师，还有参与教学工作的学院老师、团队研究生、历届本科生及企业界朋友。

今年适逢联合国宣布2015年为"光和光基技术国际年（简称国际光年）"，以纪念"光学之父"伊本·海赛姆的光学著作诞生一千年及光科学历史上一系列重要发现，借这套书的出版，向国际光年致敬！

任课教师
2015年12月15日

院长致辞 Dean address

  亲爱的同学们、亲爱的老师们、尊敬的各位嘉宾大家晚上好。"流光溢彩，魅影动人"一年一度的建筑光学作业汇报展今年恰逢国际光年，因为有了郝老师团队的十分热情，同学们便以十二分的光彩回报。因为有了诸多公司和个人的鼎力相助，同学们的光谱变得更加宽广。因为有了远道而来的各位嘉宾的观赏，同学们年轻的身影变得更加迷人。霓裳羽衣、彩云追月，最美的光影在今晚，就在今晚！你们的作品是最美的设计，你们是我心中最美的学生。

同济大学建筑与城市规划学院院长
2015 年 5 月 20 日

指导教师致辞 Tutor address

  大家晚上好，今天是同济大学光环境实验室 2015 国际光年系列活动最后一场，也是 2013 级建筑学专业"建筑物理（光环境）光影构成"作业评审展示。今晚的主题是"Let's be your light"。我们的作业素材选自音乐剧大师安德鲁·劳埃德·韦伯的代表作《剧院魅影》，通过光与音乐的互动，以光影构成的方式对其进行诠释，向 2015 国际光年致敬。我们要求学生了解新型光源——LED 的发光特点及光学特性，结合生活中常见的材料，尝试光、材质及形态的创新设计。教学环节包括课堂基础知识讲授、实验室体验教学、LED 焊接等。探索材料、光线及环境的相互关系，发挥大胆的想象，将光幻化为音乐剧中的灵魂舞者，呈现一场独特的视听盛宴。

  我们今晚活动的技术支持是上海 CREE 光电发展有限公司、欧司朗（中国）照明有限公司。同学们你们准备好了吗？让我们开始吧！

任课教师
2015 年 5 月 20 日

教学花絮 Teaching highlights

规划系的王伟强教授、景观学系刘悦来副教授、建筑系章明教授，他们都非常热心于"光影构成"教学，2015年还特别邀请了学院研工办李疏贝老师作为艺术指导加入整体的策划。

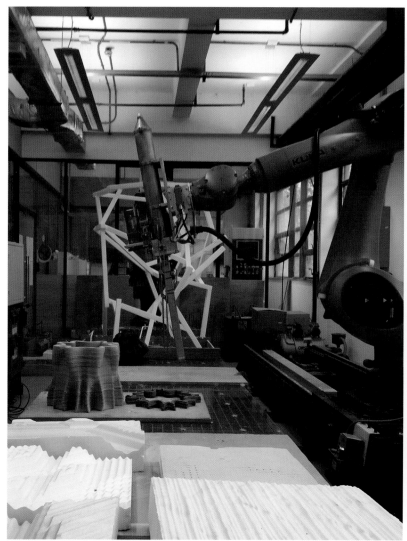

学院数字设计研究中心臧伟老师专为此次教学制作了流光魅影的主题面具标识。该面具标识运用逆向工程技术，采用三维激光扫描建立网格面，最终用五轴数控加工机床完成。

右下图是上海同济天地创意设计有限公司孙大旺所长专为此次教学汇报展制作的"国际光年"纪念徽章和"流光魅影"纪念徽章。

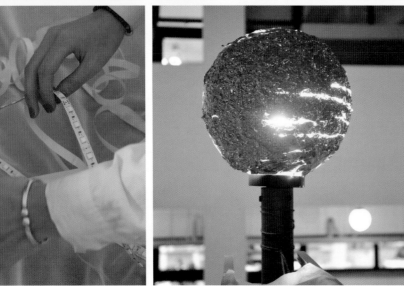

作业简介 Task introduction

一、作业题目

流光魅影 "Let's be your light"

二、作业目的

亲自动手制作光艺术装置，利用光影色彩来再现音乐剧《剧院魅影》中的经典片段，通过音乐与光的互动展现戏剧场景的情调与氛围，感受光艺术的魅力，并了解低碳环保的新型光源——LED 的发光特点、光学特性，进行 LED 焊接等实践操作，结合生活中常见的材料，尝试光、材质及形态的创新设计。观察各种材料与光线及环境的互动关系，发挥大胆和丰富的想象，将光幻化成音乐剧中的灵魂舞者，呈现一场独特的视听盛宴。致敬 2015 国际光年。

三、作业内容与方法

1. 使用材料

光源：以 LED 颗粒为主（实验室提供）共 10 种不同光色的单色 LED 芯片以及 RGB LED 芯片。
其他材料不限（各组成员自备）。推荐使用生活过程中产生的废物，如啤酒瓶、易拉罐、矿泉水瓶、方便面袋、垃圾袋等，循环利用，力求对环境影响的最小化。

2. 方法

LED 焊接：利用电烙铁等工具将 LED 芯片与电线焊接。
装置设计：探索光与材料的特性，进行光艺术效果实验，选择符合主题的效果。

3. 要求

以各组抽签选定的《剧院魅影》唱段为主题，将音乐剧片段的情节、情感、氛围融入光艺术装置中（装置体积的大小要能使所有观众席获得清晰的视看效果，装置不要太重且具有可运输性、可拆卸性、安全性），在红楼钟庭进行表演展示。

四、作业组织

以班为单位,每班分成 2 组,共 10 个工作小组,每组 15-16 名学生。
各组所表现的音乐剧片段由抽签决定。

五、成果提交

6 页设计成果文本(模版提供)。
现场成果展示。

六、作业成绩评定

成果展示当日进行公开评审。
最终成绩根据展示效果和每个人课程参与度综合评定。

七、时间安排

作品展示:2015 年 5 月 23 号(周六)19:00-20:00
展示地点:建筑学院 B 楼钟庭
评审颁奖:建筑学院 B 楼钟庭 20:00-21:00

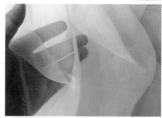

前期脚本策划

开场前音乐：Emtr'acte（QQ 音乐精编版）

时间点：0'00"00-0'02"00
背景音乐：Overture（特别版 + 原声带）
场景内容：吊灯缓缓升起，音乐点亮舞台灯火。场景闪回至1881年，歌剧院热烈和气派的场面展现在眼前。
色调和氛围预设：宏大、华丽，暖色调。
备注：表现吊灯升起点亮，与舞美灯光配合开场。

时间点：0'02"05-0'04"05
背景音乐：Think of me（电台广播版）
场景内容：正式演出中，女主角独唱技惊四座。
色调和氛围预设：暖色调，女主角为焦点。
备注：单人演出场景，周边有观众。

时间点：0'04"10-0'04"15
背景音乐：Angle of music（1986 年 London 版）
场景内容：女主角与好友合唱，梦想与现实中徘徊的少女的内心独白。
色调和氛围预设：柔美、亲切、少女感，较暗的暖色调。
备注：双人演出场景，轻柔的内心独白。

时间点：0'04"20-0'07"20
背景音乐：Music of night（1986 年 London 版）
场景内容：地下室，男女主角在一条船上，男主角低沉温柔倾诉心中情感，声音越来越低沉，在夜色中湮没。
色调和氛围预设：冷色调，渲染地下室幽暗阴冷。
备注：表现地下室场景，迷雾中蜡烛幽明闪烁，舞台灯光可提供水波纹灯。

时间点：0'07"25-0'11"28
背景音乐：All I ask of you（1986 年 London 版）
场景内容：女主角和男二号逃到歌剧院屋顶，整个巴黎尽收眼底，二人倾诉爱慕之情。
色调和氛围预设：夜色、浪漫，冷色调。
备注：天台场景，画面由动转静，男女合唱，可用追光灯。

时间点：0'11"33-0'14"33
背景音乐：Wish you were somehow here again（特别版+原声带）
场景内容：下着雪的清晨，女主角感到彷徨无助，跑到了父亲的墓园，演唱这首歌。
色调和氛围预设：凄凉、悲伤，冷色调。
备注：女主角独唱，表现墓地肃杀凄凉的气氛。

时间点：0'14"38-0'17"38
背景音乐：Wandering child（剪辑后，25周年纪念版）
场景内容：魅影出现在墓地，召唤女主角一起离开，男二号出现。
色调和氛围预设：冷色调，凄凉、悲伤。
备注：男女主角对唱，男主在高处，可用追光灯。

时间点：0'17"43-0'20"43
背景音乐：The point of no return（电台广播版）
场景内容：《唐璜》首演当晚魅影代替原来主角与女主角合唱。
色调和氛围预设：暖色调，小场景。
备注：男女主角对唱，渲染紧张气氛，此处为全剧情感最高潮。

时间点：0'20"48-0'23"48
背景音乐：Finale（电台广播版）
场景内容：女主角和男二号离开，男主角在地下室独自一人，内心悲怆凄凉。
色调和氛围预设：冷色调，音乐旋律清新。
备注：男主角独唱。

时间点：0'23"53-0'26"53
背景音乐：Masquerade（电台版）
场景内容：化妆舞会、一群人穿着奇装异服欢乐歌唱。
色调和氛围预设：暖色调，可加彩色灯光，场面宏大热闹欢快。
备注：第十组先表演，结束时灯光照亮全场，所有人一起歌唱舞动，最后面具亮相。

谢幕音乐：Emtr'acte（QQ音乐精编版）

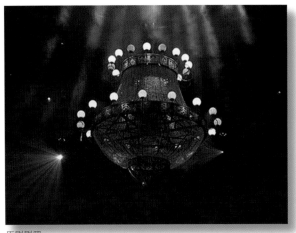

原剧剧照

[Overture]

1911 年，歌剧院中"魅影事件"遗留物的拍卖会现场。

最后一件拍品是在著名的"魅影事件"中摔得支离破碎的枝形大吊灯 (Chandelier)。

拍卖师为吊灯安装上新的电灯，向众人展示其昔日风采，随着电路接通，乐队和管风琴奏响恢弘的序曲，华丽的吊灯缓缓升起，发出耀眼的光芒。音乐重新点亮了已经破败的歌剧院，场景切回到1881年。

原剧剧照

[Think of me]

原本演侍女的舞蹈演员克莉丝汀（女主角）被人推选出来代替原来女主角卡罗塔。彩排中完全不被新经理看好的克莉丝汀技惊四座，这时的钢琴伴奏逐渐切换成管弦乐同奏，彩排也变成了正式演出。克莉丝汀的演出大获成功，美妙的歌喉征服了全场观众，包括二楼包厢里的剧院赞助人——贵族罗尔（男二号）。

原剧剧照

[Angel of music]

克莉丝汀向挚友梅格自述说,她的演唱能力得益于一个神秘的老师,他似乎无处不在,却从未真正出现在面前,只是不断用自己的声音引导她,就像父亲临终前所说的"音乐天使"那样令自己向往。此时罗尔也来祝贺克莉丝汀,并坚持邀请她共进晚餐。当克莉丝汀回到化妆室时,房间里却响起了魅影(男主角)的声音。

原剧剧照

[Music of night]

歌剧院的地下湖,克莉丝汀与魅影坐在一条船上,蜡烛在迷雾中闪烁着幽明的光,似有若无的风在黑暗中吹动,他们唱起了夜晚的音乐,歌声柔美动人。魅影带着克莉丝汀来到了地下迷宫的深处,这里就仿佛他的王国一般。伴随魅影美妙的歌声,克莉丝汀沉浸在梦幻般的音乐世界。

原剧剧照

[All I ask of you]

被套索绞死的控幕员布克的尸体从舞台天幕上掉下来,观众们尖叫着逃离。罗尔赶来将恐惧的克莉丝汀带离舞台,两人仓促之中躲到了剧院天台上。罗尔表示愿倾尽全力来保护克莉丝汀,克莉丝汀深受感动,答应了罗尔的表白,愿与他厮守,两人沉浸在对爱情的美好希冀中。

原剧剧照

[Wishing you were somehow here again]

为了避开魅影控制,克莉丝汀被罗尔藏了起来。午夜,克莉丝汀在睡梦中起身,偷偷前去自己父亲坟墓前祈祷、倾诉、寻求解脱。在墓碑前,她祝愿可怜的魅影也能得到幸福,她祈祷父亲能指引自己。

原剧剧照

[Wandering child]

魅影伪装成克莉丝汀父亲的声音,用一曲极具魅惑力的"Wandering child"开始引诱她,克莉丝汀仿佛被父亲的影子所吸引,开始情不自禁地走向魅影……

[The point of no return]

原剧剧照

《唐璜》演出当晚,魅影已经神不知鬼不觉地代替了原来的男主角,出演带着面具伪装成好友去戏弄对方妻子的唐璜。只有克莉丝汀敏感地察觉到与自己合演的正是魅影本人,两人假戏真唱,一曲"The point of no return"将剑拔弩张的气氛推向高潮。在两人的较量中,再一次仿佛被感动的魅影似乎失去了置克莉丝汀于死地的敌意。在魅影悲壮的歌声中,克莉丝汀鼓起勇气,在众目睽睽之下将他的面具脱下……

原剧剧照

[Finale]

魅影面对心爱的八音盒，独自吟唱起象征告别、毁灭和新生的咏叹调。众人终于来到了这里，却已人去楼空，只在空荡荡的大厅里发现了一个面具……从此，再也没人听过魅影那魔幻的声音……

原剧剧照

[Masquerade]

剧院举行着盛大的化妆舞会，穿着奇装异服的人们打扮成魅影的样子合唱华丽的乐曲"Masquerade"，庆祝和嘲笑不攻自破的对手。偷偷订婚的克莉丝汀与罗尔也在人群中。众人正在狂欢的最高潮，化妆成死神的魅影突然出现。

彩排现场 Rehearsal

2015年5月13日初次分组彩排 First group rehearsal

2015年5月15日总体彩排 Overall rehearsal

2015年5月23日 汇报准备 Preparation for the final report

TRACK LISTING
场景展示顺序

开场致辞

1 Overture 2'45"

姚冠杰（组长） 黄舒弈（副组长） 石文鑫 窦嘉伟 周雨茜 熊非 花炜 侯苗苗 成紫玙 何侃轩 杨挺 冯羽奇 田园 周博 董越斌

2 Think of me 2'14"

孙少白（组长） 申程（副组长） 贾姗姗 陆亦宁 龙嘉雨 孙一诺 熊晏婷 张晓雅 朱玉 罗辛宁 周锡晖 鲁昊霏 王劲扬

居子玥 王梓安

3 Angel of music 2'11"

乔映荷（组长） 陈有菲（副组长） 吴庸欢 徐幸杰 黄靖 邓浩彬 郝应齐 黄娜伟 蔡庆瑜 张靖 时任夏莹 杨鹏宇 申朝阳

张谨奕 郭秋花

4 Music of night 3'08"

谢越（组长） 马嫣砾（副组长） 潘思雨 徐思璐 雒雨 周易 苏家慧 王竹韵 朱尽染 王萧迪 蔡东旭 李安贵 周与锋 卡琳娜

郭绵沅津

5 All I ask of you 4'14"

郑海凡（组长） 孙正宇（副组长） 胡伟林 甘崇雨 徐琛 林敏薇 吴璇 江旭莹 许纯 路秀洁 解远志 尹铉玟 林辰彻

6 Wish you were somehow here again 3'05"

焦威（组长） 徐语键（副组长） 唐倩倩 岑丹莺 张昊胄 代昊辰 Dana 蔡毓怡 李昊 张溯之 张晶轩 常馨之 高舟 吴雨 Tim

7 Wandering child 2'55"

郑少凡（组长） 赵启（副组长） 张煜 王江辉 方志浩 覃杨 邸文博 王子若 揭晶皓 吴逸南 李想 张玉娇 周顺宏

李淑一 别雨璇

8 The point of no return 2'36"

高子晗（组长） 郭皓阳（副组长） 任宇桓 张佳择 段睿妍 倪凌 翁子健 唐垲鑫 胡艺龄 周雨桐 周逸昀 张倩

李英子 彭心悦 李宛蓉 朱冰洁

9 Finale 3'

金世煜（组长） 王兆一（副组长） 聂方达 陈俐 冯田 陈路平 郝明宇 张万霖 庄铭予 何斌贤 王旭东 高雨辰 谢成龙

唐奕诚 刘雨婷 卡那库亚

10 Masquerade 3'56"

和亦宁（组长） 孙晓梦（副组长） 肖佳蓉 杜昆蓉 王妍 朱任杰 林亦晖 云贺 金刚 张晚欣 张文雪 张治宇 郑国臻 陈宇宁

安以静

评审颁奖

整场活动时长约为40分钟

作品展示汇报 Final showcase

任课教师及作业策划：郝洛西 教授

背景音乐：The phantom of opera
《剧院魅影》
安德鲁·劳伊德·韦伯

组　　长：黄舒奕　姚冠杰
　　　　　孙少白　申　程
　　　　　乔映荷　陈有菲
　　　　　谢　越　马嫣砾
　　　　　郑海凡　孙正宇
　　　　　焦　威　徐语键
　　　　　郑少凡　赵　启
　　　　　高子晗　郭皓阳
　　　　　王兆一　金世煜
　　　　　和亦宁　孙晓梦

展示时间：2015 年 5 月 23 日　20:30-21:10
展示地点：同济大学建筑与城市规划学院
　　　　　B 楼钟庭

开场序幕 Prelude

团队硕士研究生张萌和付美祺负责现场引导

院长李振宇教授为支持教育的照明企业颁发荣誉证书

第一幕
光轮 Overture

为了符合"剧院魅影"的主题,第一幕的灯装置需要营造出恢弘中带着点诡秘的开场气氛,大家讨论决定采用歌剧里大吊灯的形式,并且突出效果震撼、变化丰富、经济环保三个特点。首先我们把黄色滤光纸贴在回收的矿泉水瓶上,再套在 LED 灯上,制成主灯。接着将这些灯固定在定制的灯架上。然后采用白色柔性 LED 灯带、草帽 LED 灯串等装饰以增添大吊灯的气势。我们还设计出了安全且便于操控的电路以配合音乐脚本创造丰富的灯光变化。

Our lighting needs to creat spectacular and a little mysterious atmosphere of overtrue to fit the themes of the Phantom of the Opera. After discussions, we decided to use the form of the chandelier, and utilize its easy&economic, powerful and variable characteristics. Our main lamp has four circles from top to bottom. Each of the lamp consists of a recycling water bottle pasted with yellow filter paper and an LED. Then we install white flexible LED strips, LED stars and various kinds of lamps to make the chandelier more gorgeous. Also we design safe and easily controllable curcuits to coordinate with music script.

作者 Designers

姚冠杰　黄舒弈　花炜　成紫玗　冯羽奇　杨挺　熊非　田园
侯苗苗　周博　周雨茜　窦嘉伟　董越斌　何侃轩　石文鑫

分镜头脚本设计 Storyboarding design

时间	节点	灯光变化
	[一] 幕布下降开始 00:00–00:13	
00:00	声音出现	H 全亮 E2 亮 K 亮
00:00–00:04	第一个低音出现	E1 亮起
00:04–00:08	到第一个高音的过程	四个人跟着四个音的节奏做出相应动作，并缓
00:08–00:11	到第二个低音出现	缓下蹲到半蹲状态
	第二个低音出现	HK 灭掉
00:13	鼓点	HF 闪两下
	[二] 乐器进入 推动	
00:13–00:31		灯的上升过程
		追光 F 追灯
00:13–00:19 强	保持的高音开始时	H1 H2 每拍两下依次亮起
00:19–00:23 弱	保持的低音开始时	H3 H4 每拍两下依次亮起
		H1 H2 灭
00:23–00:27 强	保持的高音开始时	H1 H2 每拍两下依次亮起
00:27–00:31 弱	保持的低音开始时	H3 H4 每拍两下依次亮起
鼓点	暗	H3 H4 每拍两下依次亮起
	[三] 细碎节奏	
00:31–00:38		H1 H2 H3 H4 交替乱闪
	[四] 低音乐器进 00:38–00:53	
00:38–00:46	三个音	H1 H2 H3 H4 交替亮起
	（小节奏）	K1 K2 亮亮灭灭亮亮灭灭
	高音	H1 H2 H3 H4 一起亮
	两个音	H1 H2 H3 H4 逐次灭掉
	低音	H 亮
00:46–00:53	三个音	H1 H2 H3 H4 交替亮起
	高音	H1 H2 H3 H4 一起亮
	两个音	H1 H2 H3 H4 逐次灭掉
	低音	H 亮
	[五] 音调提高 00:53–01:10	
1:03	开始四个音	第二个 H1 亮 第三个 H2 亮 第四个 G 亮
	八个音调下降的音	两两一组 DCBA 分别闪烁
	[六] 桥 01:10–01:17	
01:11		J 亮
01:12		G 灭
01:13	01:11 是第二个重复	F 灭
01:14		H 灭
01:15		J 灭
01:16 全灭	最后一个音	全灭

时间	节点	灯光变化
	[七] 高潮 01:17–01:48	
	第一个重音	AB 亮
	第一个重音	AB 亮
01:18	第二个重音	CD 亮
	第三个重音	ABCD 灭
	第四个重音	除 J 全亮
01:20–01:28		保持除 J 全亮
01:28	长音	ABCD 灭
	第一个重音	A 亮
01:25	第二个重音	B 亮
	第三个重音	C 亮
	第四个重音	D 亮
01:28–01:35	01:35 长音	除 H 灭掉
	第一个重音	A 亮
01:33	第二个重音	B 亮
	第三个重音	C 亮
	第四个重音	D 亮
01:38–01:44		保持
01:44	八个音调下降的音	两两一组 DCBA 分别闪烁
	[八] 坠落 01:48–01:53	
01:50	重音	灯全亮 同时灯掉
	[九] 重生 01:54–02:10	
	音乐起时	追光 F 打灯上
		K 亮
02:01	铁链声	绳子往上收
	铁链声完	K 灭
	[十] 二次高潮结束 02:11–03:00	
02:12		除 K 全亮
		空中氛围 G 亮 追光扫两面墙
02:12–02:20		保持全亮
02:18	第一个重音	A 亮
	第二个重音	B 亮
	第三个重音	C 亮
	第四个重音	D 亮
02:20–02:26	01:35 长音	除 H 灭掉
02:27–02:30		乱闪
02:30–02:36	长音	所有灯保持亮的状态
02:37	八个音调下降的音	两两一组 DCBA 分别闪烁
02:45	长音	全亮
02:45–02:51	五个降音	ABCDJ 灭
02:51–02:55	五个升音	ABCDJ 亮
02:55–02:58	五个降音	ABCDJ 灭
02:59	最后一个音	全亮

注：A 最下面一圈　B 第二圈（两种 B1B2）　C 第三圈（两种 C1C2）　D 最上面一圈　E 地下氛围
F 追光　G 空中氛围灯　H 上层 LED 灯条　J 天空　K 蓝光

设计草图 Sketches

 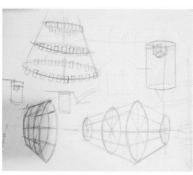 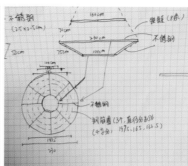 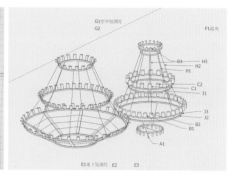

材料构成 Material

制作过程 Construction

现场安装 Installation

分镜头演出效果 Storyboarding

第二幕
漩光 Think of me

此次光装置作业的主题是结合音乐,在建筑空间中塑造变换的场景。我们组的音乐"Think of me"是整个音乐剧中少有的以温馨、纯洁的爱为主题的片段。该片段表现的是克莉丝汀舞台首秀,涉及从排练到正式演出的空间转换。因而通过屏风的展开,屏风上灯的变化塑造舞台空间。同时在场地平台的矮墙上布置 LED、镂空灯盒营造周围空间氛围。灯光以黄色和蓝色为主,色调纯净而又温暖。根据音乐旋律,我们设置了围绕前后两个高潮展开的编排,表现女主角的绽放,同时灯光根据旋律进行分组变化,与音乐巧妙结合。

The theme of the lighting is to combine music with space in order to create different scenes. Music of our group "Think of me" is a warm, pure love-themed piece in the musical. Meanwhile, the piece describes the first show of Christine, from rehearsal to the live stage. Thus we have the screen unfolding, light varying. And on the walls and stairs, we set LED and lamp box to create the desired atmosphere. Dominated by yellow and blue colors, the tone is pure and warm. Lights are grouped according to the melody changes, and music combined. So that we see the progress of a girl in her musical life, like a blooming flower, in a way of illumination arranging.

作者 Designers

孙少白　申程　龙嘉雨　王梓安　贾姗姗　张晓雅　罗辛宁　王劲扬
孙一诺　居子玥　周锡晖　熊晏婷　陆奕宁　朱玉　鲁昊霏

分镜头脚本设计 Storyboarding design

场景变换		灯光方案
第一阶段——铺垫，营造静谧的场景		
幕间		追光打在三楼末端台阶开始处 2-3s 再熄灭
0-6"	高跟鞋声音	蓝色光逐级照亮踏面
9"-17"	女配一剪影出现于幕布之后，进行舞蹈表演	蓝色光打在幕布上形成剪影
17"-25"	女配二剪影出现于幕布之后，进行舞蹈表演	白色光打在幕布上形成剪影
25"-38"	女配一与女配二舞蹈	舞台左右两端蓝白两色剪影（Ⅰ）
38"-42"	女主剪影出现并重合	红、蓝两投射灯由两侧打向女主并逐渐相中靠拢、重合（Ⅱ）
第二阶段——绽放，展现华丽的舞台		
42"	女主裙上灯亮，开始旋转 屏风灯亮，两边打开，前布落 女配灯灭	女主自控裙灯变化 屏风上黄色灯亮（Ⅲ）
42"-59"	女主旋转	女主自控裙灯变化
59"-1'17"	女主继续旋转（灯变暖） 女配裙灯亮，配合旋转 屏风灯渐暗（不灭）	女主自控裙灯变化 女配自控裙灯变化
1'17"-1'58"	女主（蓝灯暖白等同开）指挥小盒子变化，配合适当舞蹈 小盒子投射菱形光斑 屏风上四组灯呼应小盒子变化	4组小盒分别亮灯，暖色为主——红、黄、蓝、白 4组屏风灯颜色：黄、琥珀、橙、红（Ⅳ）
1'58"-2'07"	灯渐渐全灭	
第三阶段——高潮，进行惊艳的谢幕		
2'07"-2'15"	所有的灯打开	女主自控 暖色大吊灯1组控 现场光配合加亮80%（Ⅴ）
谢幕		音乐闭 灯光渐暗

设计草图 Sketches

材料构成 Material

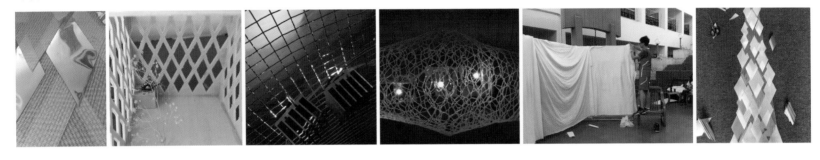

制作过程 Construction

现场安装 Installation

分镜头演出效果
Storyboarding

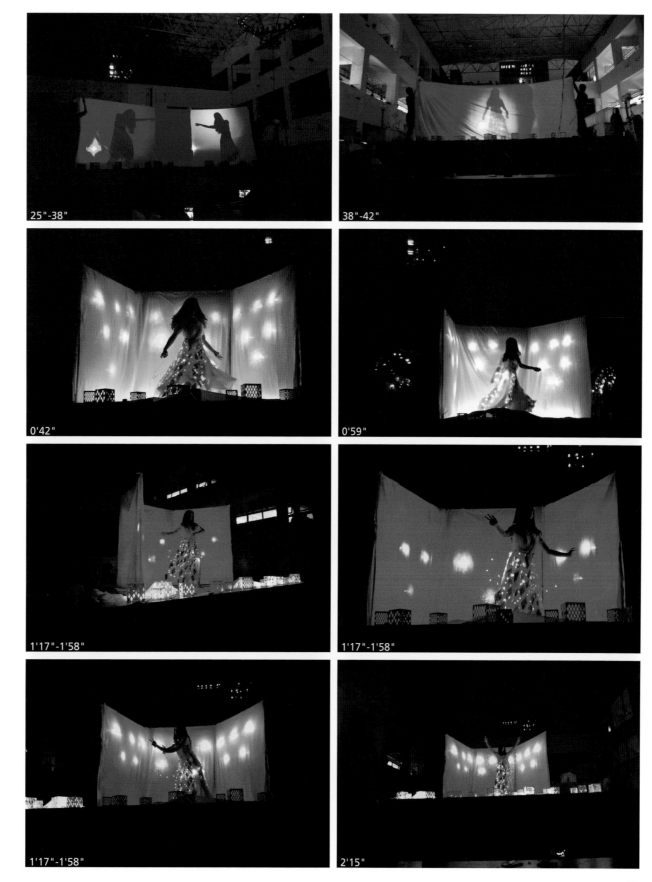

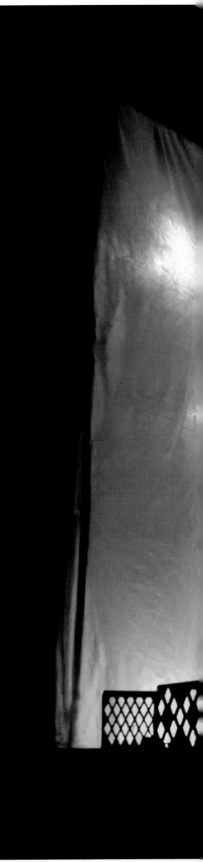

第三幕
冰影雾缭 Angel of music

我们的场地是一个长长的、曲折的、有槽的平台，根据这个条件，我们让凹槽成为一条光源，打出灯光映射在幕布上，我们的装置依托平台向上升起，成为跳跃的彩色体块。

首先我们在凹槽中摆放 LED，再跨着凹槽架起一排拷贝纸，这样就有了承影面。然后我们在体块内放置贴了锡纸的刻花纸杯，这样拷贝纸上就能呈现出纸杯上的花和锡纸反射的灯光。另外，平台上和高处气球的灯光让装置的图案和色彩更加丰富。

Our site is a long and tortuous platform which has a notch on it. On this condition, we decided to utilize the notch as a light source, to bring out the light on the screen. Our installation is rised up on the platform to be a color block.

Firstly, we put LEDs in the notch, then we surround the notch with a layer of paper and through that we got a screen. Secondly, some cups with hollow have been put into the block, then we got light flower on the screen. Besides, other light on the platform and balloons with water on the frame make the pattern and color of the device more fabulous.

作者 Designers

蔡庆瑜　陈有菲　邓浩彬　郝应齐　黄　靖　黄娜伟　乔映荷　申朝阳
时任夏莹　吴庸欢　徐幸杰　杨鹏宇　张谨奕　张　靖　郭秋花

分镜头脚本设计 Storyboarding design

第四槽						第三槽				第二槽						第一槽		
蓝黄	黄黄	蓝黄	黄黄	蓝黄	蓝黄	绿绿	蓝蓝	绿绿	白白	白红	白蓝	白红	白白	白红	白蓝	白白	黄黄	白白
4	3	3	2	2	1	2	1	1	4	4	2	3	3	2	1	2	1	1
绿	黄	绿	黄	白	白	蓝		绿	白	绿	蓝	白	蓝	红	白	红	白	红	蓝	黄	白	黄	白	蓝
4	4	4	3	8	8	7		7	2	6	1	6	5	5	4	3	4	2	2	1	3	2	2	1	2	1	1
	黄		白			蓝				绿	绿	绿	黄	黄		白				蓝		黄		蓝	白	蓝
	4		4			4				3	2	1	3	2		3				2		3		1	2	1

时间	音乐状况	灯位	详细
0:00–0:18	女二起奏	槽灯	蓝色氛围③⑧④⑤②⑦①⑥
	四句歌词		0:18一起收回（分八组）
0:19–0:29	小提琴纯乐	挂灯B	1、1、1、2、2、3、3、1、1
			随重音打开（8个重音）
0:30–0:50	女主倾诉	槽灯A	体快感初现，蓝+白+暖白
	悠扬	挂灯B	杯子转动：白+黄（单色）
0:51–1:10	女主	槽灯A	提高亮度+黄
	激动	挂灯B	转动，颜色混合
		台灯C	从后往前亮
1:11–1:21	女二	槽灯A	亮单色，蓝+白
	激动	挂灯B	变单色
		台灯C	关掉2/3

时间	音乐状况	灯位	详细
1:22–1:29	女二	同上A	白+暖白
	激动	B	混色
		C	不变
1:30–1:51	合唱	A	蓝+白+暖白+黄+绿
	高潮	B	混色，转动
		C	全亮
		柱灯D	一齐亮，1:30/1:51灭掉
1:52–2:11	音乐渐变	A	一绿一黄一白，留蓝，然后灭
		B	单色，留到最后
		C	从后往前灭

设计草图 Sketches

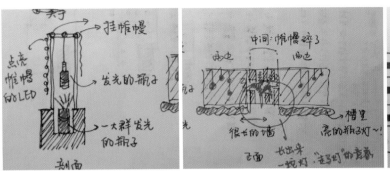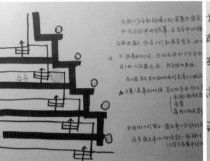

材料构成 Material

制作过程 Construction

现场安装 Installation

分镜头演出效果
Storyboarding

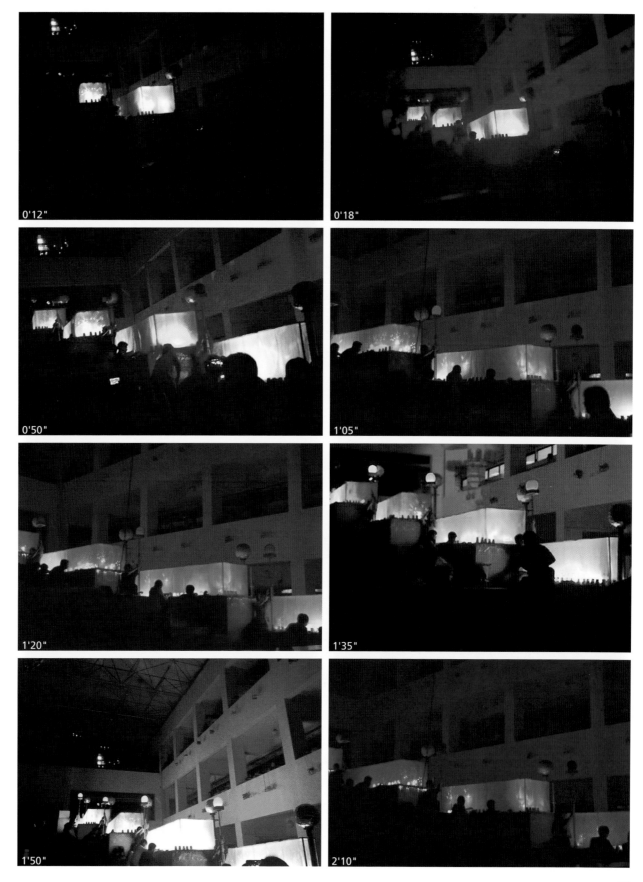

第四幕
夜辉律动 Music of night

我们组的音乐描绘了地下室中有水有烛光的一幕情景。根据环境背景我们决定通过装置来展现夜幕与水。我们用铝箔纸串起珠帘，通过对其底部灯光的反射和在天顶形成的影子，表现夜的星光点点。又用装有半瓶水的矿泉水瓶组成多种颜色 LED 灯装置来表现摇曳的水波。

"Music of night" is a drama performed in the basement by water with candlelight. According to this background, we decided to show the night and water by means of the device. We chose aluminum paper and line to make curtains. Through reflection at the bottom of the polish and at the zenith of shadow, night light scattered. Several bottles of half water were recombined with LEDs to show the swaying of the water.

作者 Designers

谢越　周易　王竹韵　朱尽染　马嫣砾　徐思璐　潘思雨　雒雨
周与锋　郭绵沅津　王箫迪　李安贵　蔡东旭　苏家慧　卡琳娜

分镜头脚本设计 Storyboarding design

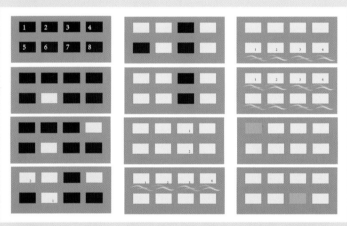

1. 本段主要是一个音乐引入过程，主要运用第一层帘灯，同时亮起一些水灯，为之后做铺垫。

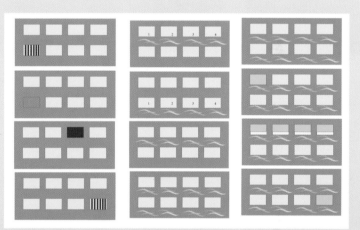

2. 此段是副歌第二遍，主要运用"水灯"，同时配合第二次"帘灯"以暗示。

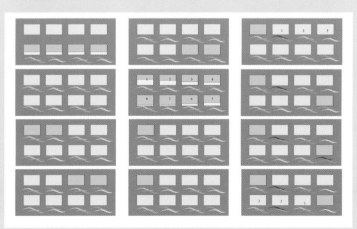

3. 本段主要是水灯的冷暖变化配合音乐节奏，八个窗口依次切换。

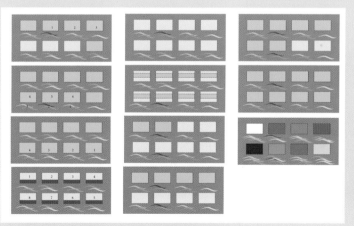

4. 结尾部分，随着音乐进入高潮，灯光亮度增加，变化加多，颜色更为丰富。

利用本组场地内的八个窗口及与其相连接的墙面，设置八组装置，每组分别为"帘灯"和"水灯"，各有一名同学管理开关，"帘灯"有三种不同的亮度，"水灯"有冷暖两种颜色组合，根据音乐节奏与旋律进行编排，同时整体形成从暗到亮的效果。

设计草图 Sketches

材料构成 Material

制作过程 Construction

现场安装 Installation

分镜头演出效果
Storyboarding

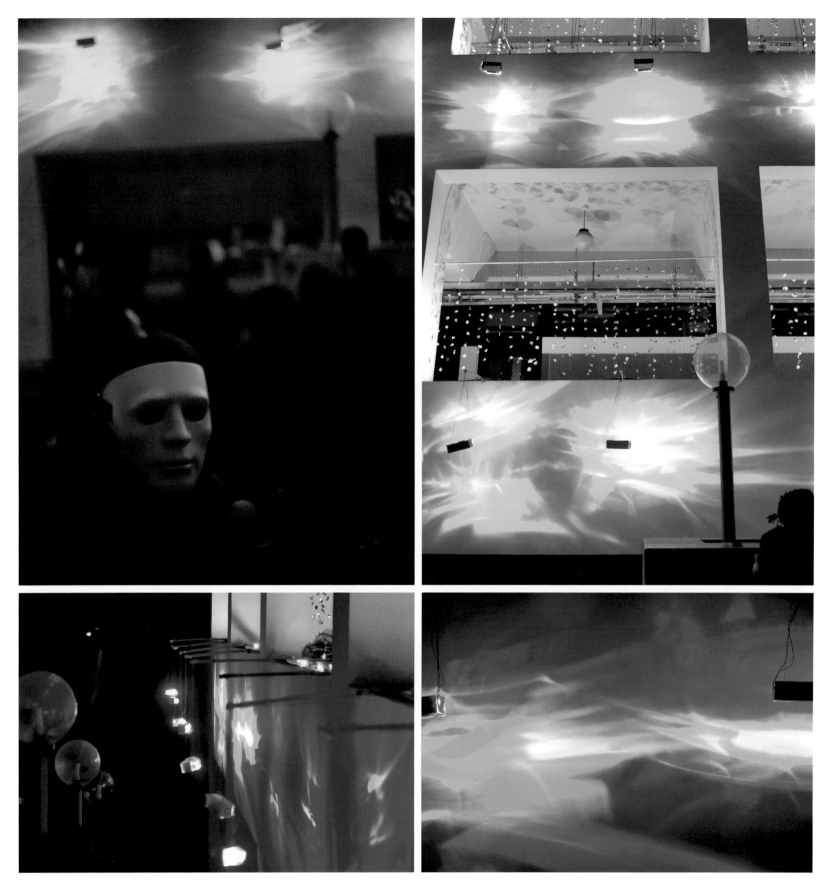

第五幕
两生花 All I ask of you

克莉丝汀与罗尔在惊慌中逃到了屋顶上，罗尔一面安慰她说这一切都是梦境，一面向克莉丝汀表示了埋藏已久的爱意，两人合唱了一曲"All I ask of You"互表爱意。这一段情节是全剧中最温馨浪漫的部分，因此我们选择运用投影作为主要的呈现方式。通过投影产生的男女主角剪影交替出现对应歌曲中男女声的唱段；红黄蓝三种色光映在白色墙面上的鲜花剪影暗示了爱情的萌芽；为了配合故事中唯美浪漫的效果，我们还运用瓶底投影的方式，渲染舞台气氛。花和人物的影像交织更替，音乐渐入高潮而终归平静。

Christine and Raoul fled to the roof in panic. Raoul comforted Christine that all this is just a dream and told her that he had fallen in love with her for a long time. Two people chorused the song "All I ask of you" to expresss their love. This episode is the most romantic part in the drama. We choose to use projection as the main presentation mode. Generated by the projection silhouettes of men and women appearing alternately, it corresponded with the song in male and female voice of the aria. Red, yellow and blue lights formed the silhouette of flowers on the white wall, when suggests the seeds of love. In order to cope with the beautiful and romantic effect of the story, we also use bottle bottom projection, rendering the atmosphere of the stage. The replacement and interwoven of the flowers and figures enter a climax and return to peace with the floating of music.

作者 Designers

郑海凡　徐琛　甘崇雨　解远志　吴璇　尹铉玟　林辰彻　林敏薇
胡伟林　许纯　路秀洁　孙正宇　江旭莹

分镜头脚本设计 Storyboarding design

场景号	时间	草图	编排内容	灯光变化
1	0'00"–0'13"	草图 1	故事开始，场景引入	地灯从上到下依次打开，模仿脚步的节奏。
2	0'13"–0'48"	草图 2	男主从上向下走与女主汇合，男主独唱	男主投影灯从上至下依次打开，走过的地方开两盏灯照亮花。
3	0'48"–1'24"	草图 3	女主从下向上走迎接男主，女主独唱	女主投影灯从下至上依次打开，走过的地方开两盏灯照亮花。
4	1'24"–1'57"	草图 4	男女主角相互走进，短暂相遇，对唱	男女主投影灯间隔打开，同时打开相邻花灯。
5	1'57"–3'17"	草图 5	男女主角转身远离后又再次相互靠近	男女主投影灯一次按照近—远—近的顺序亮灯。
6	3'17"–3'45"	草图 6	男女主角在场地中心相拥	先亮中间双人灯，音乐高潮时所有华灯一起亮，之后一次熄灭。剧终。

草图 1　草图 2
草图 3　草图 4
草图 5　草图 6

设计草图 Sketches

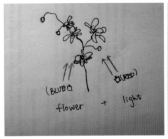 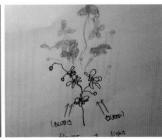  

材料构成 Material

  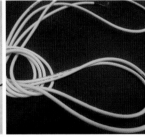   

制作过程 Construction

现场安装 Installation

 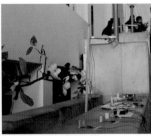   

分镜头演出效果
Storyboarding

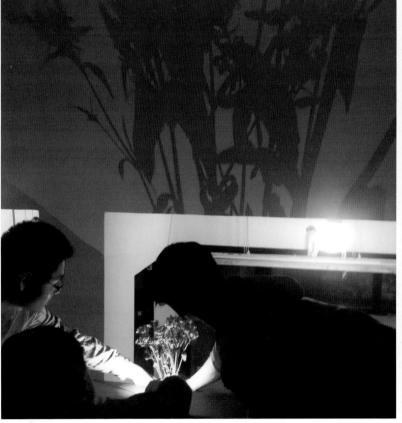

第六幕
追忆似水年华 Wishing you were somehow here again

在这一段音乐剧中,女主人公来到了父亲的墓地,想起了过去和家人度过的时光,黯然神伤。我们设计的装置采取了墓碑的形式,其上巨大的十字架形缺口,试图表现生者与彼岸的巨大鸿沟。十字架形的缺口根据人体尺度设计,演出者在狭缝中的舞蹈,讲述了女主人公失去亲人的怅惘与哀伤。我们利用不锈钢管与PVC管搭建骨架,错动地固定卡纸板,LED发出的灯光透过狭缝打在纸板上,投射出柔和的光线。表演中通过分别控制每个模块的开关,使灯光配合音乐的起伏与舞者的动作如呼吸般亮灭。

The installation forms a tomb with a cross-shape crack on it. It is designed to present the unreachable distance between the living and the dead. The cross-shape crack is designed to fit the scale of human, so the dancer's performance in the narrow space can express the sorrow of the heorine who lost her father.
We use stainless steel tubes and PVC tubes to form the frameworks. Layers of cards are pinned upon them and the warm light of LEDs shines across the slits. Every module of LEDs is controlled in concert with the music and the performance of the dancers.

作者 Designers

分镜头脚本设计 Storyboarding design

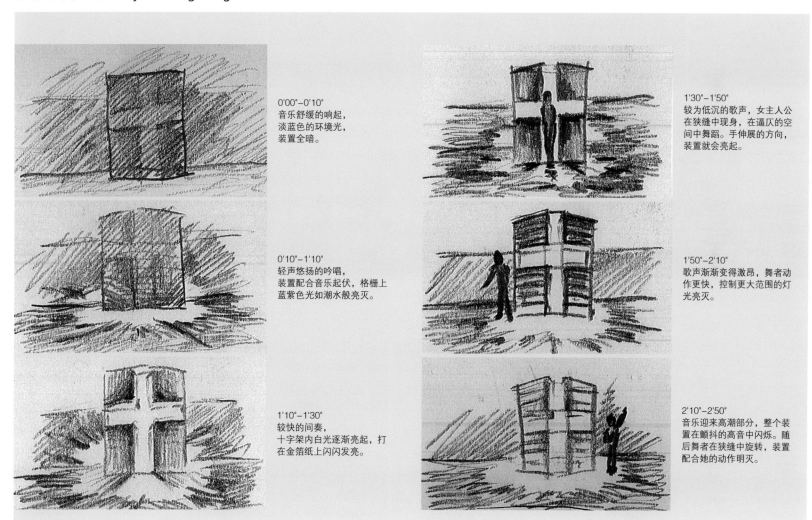

0'00"–0'10"
音乐舒缓的响起，
淡蓝色的环境光，
装置全暗。

0'10"–1'10"
轻声悠扬的吟唱，
装置配合音乐起伏，格栅上
蓝紫色光如潮水般亮灭。

1'10"–1'30"
较快的间奏，
十字架内白光逐渐亮起，打
在金箔纸上闪闪发亮。

1'30"–1'50"
较为低沉的歌声，女主人公
在狭缝中现身，在逼仄的空
间中舞蹈。手伸展的方向，
装置就会亮起。

1'50"–2'10"
歌声渐渐变得激昂，舞动
作更快，控制更大范围的灯
光亮灭。

2'10"–2'50"
音乐迎来高潮部分，整个装
置在颤抖的高音中闪烁。随
后舞者在狭缝中旋转，装置
配合她的动作明灭。

设计草图 Sketches

   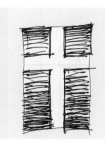 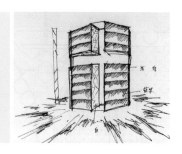

材料构成 Material

     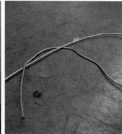

制作过程 Construction

现场安装 Installation

分镜头演出效果
Storyboarding

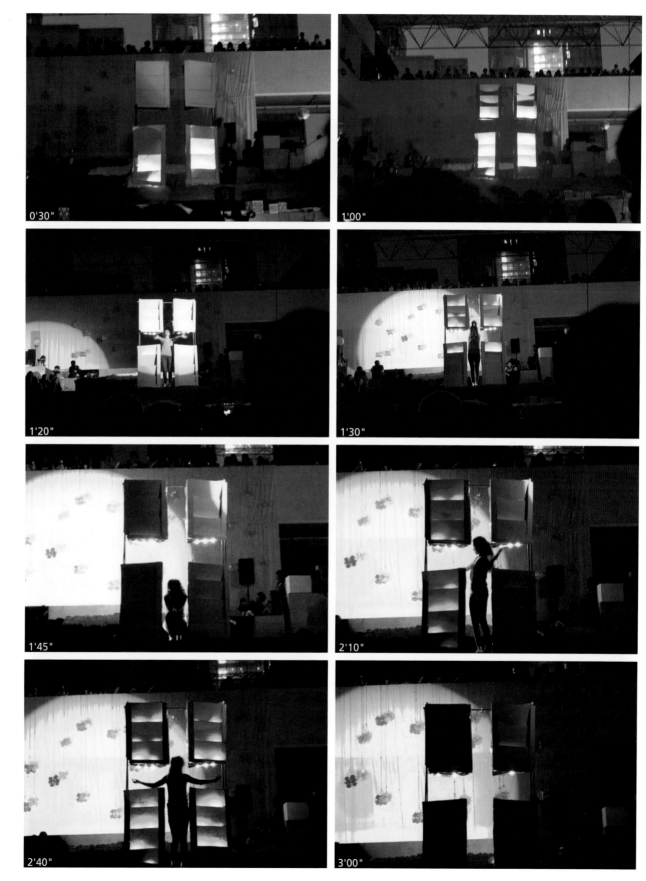

第七幕
幻影移形 Wandering child

这一幕主要讲述了女主角在墓地邂逅魅影并用歌声与魅影互动的场景。我们根据音乐将表演分成两段。第一段纯音乐部分，分别用杯子、镂空的纸筒制成三组装置，根据旋律来控制灯光亮暗，光影投射在墙面上，烘托墓地的氛围。第二段歌声响起，男女主人公穿着缝有 LED 灯的斗篷、裙子出场表演。为了表达男女主人公精神上的若即若离，我们采用了双女主角的模式，高处和低处的女主角身上的灯光装置轮番亮暗，与男主角互动，三人身上的灯光交错，体现了男女主人公的纠缠与女主角内心的纠结。

This scene focuses on the heroine in the cemetery encounter and interact with Phantom of the Opera. We divided our show into two sections according to the music. First paragraph Instrumental part, three devices consist of cups and hollow tubes. According to the melody, we controled the light and projected to the wall to heighten the atmosphere of the cemetery. The second paragraph, hero and heroine are wearing a cloak and skirt with LED lights. In order to express the ambiguous love between the hero and heroine, we have adopted a dual mode of heroine. They take turns to control the light on and off and interact with the hero, reflecting the struggle between the hero and heroine.

作者 Designers

张 煜　张玉娇　周顺宏　李淑一　别雨璇　邸文博　李 想　赵 启
吴逸南　覃 杨　王子若　郑少凡　方志浩　揭晶皓　王江辉

分镜头脚本设计 Storyboarding design

灯光变化（冷色光）。
0'11"–0'41"

灯光变化（冷色光）。
0'41"–1'24"

随音乐渐弱，墙灯逐渐熄灭。
1'24"–1'40"

中庭场地内女主1号，与二层男对唱（女主一号身上灯光亮起，男主所在窗口背景灯变化）。
1'40"–2'30"

女主消失，另一女主在男主身边突然出现，灯光配合舞蹈。
2'30"–2'56"

设计草图 Sketches

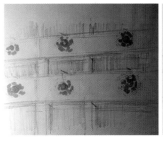 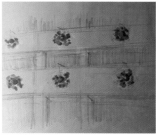   

材料构成 Material

     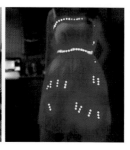

制作过程 Construction

现场安装 Installation

分镜头演出效果
Storyboarding

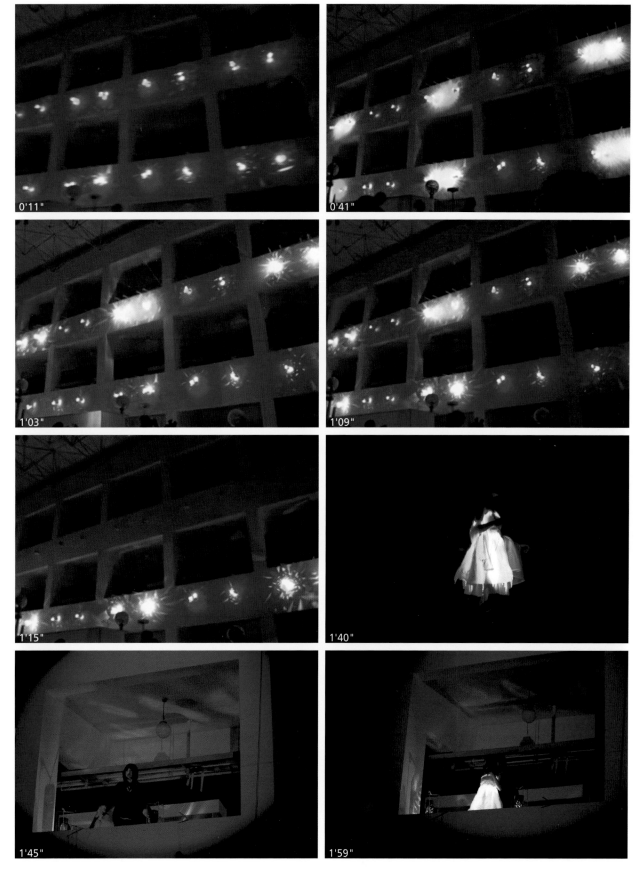

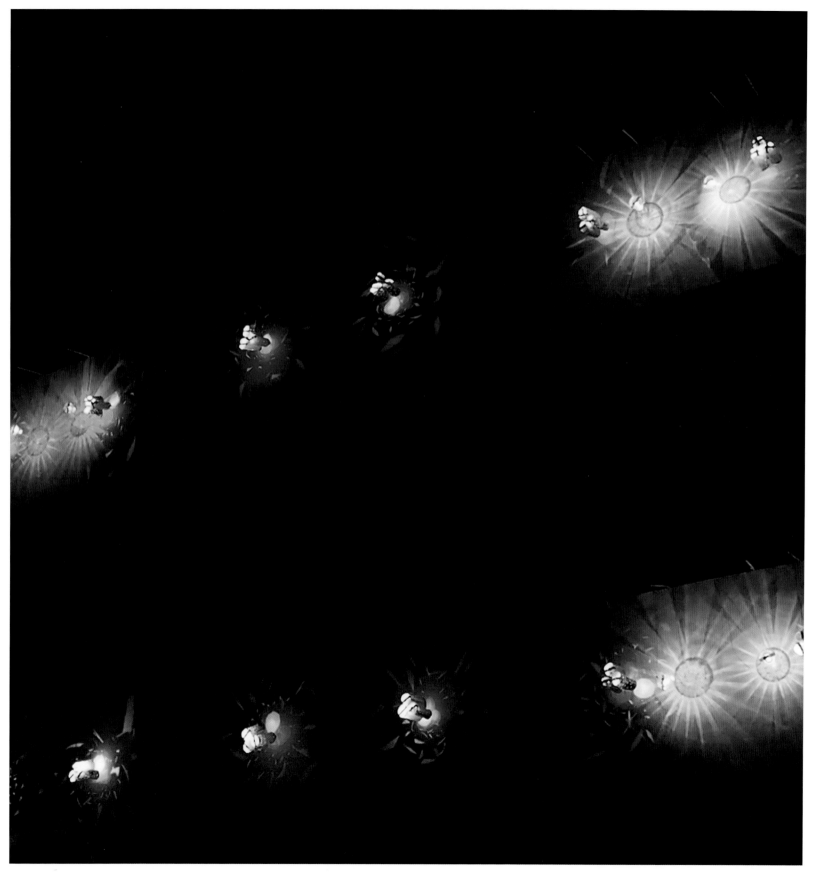

第八幕
不归路 The point of no return

此装置设计灵感源于该幕音乐剧的标题"The point of no return"。该幕为这部音乐剧的剧中剧。该剧情感在这一幕发生转折。第一段为克莉丝汀 的独唱部分,妖艳而充满活力;第二部分为魅影与克莉丝汀的合唱,缠绵悱恻,若即若离。而在最后一部分当中,《剧院魅影》的主旋律响起,魅影将面具摘下,全剧迎来最高潮。设计时,结合我们小组的场地特点,决定将灯分为蓝红两种颜色,分别代表男女主角。通过控制灯光的变化,展现男女主角的互动与情感变化。

This device is inspired by the opera's title screen——"The point of no return". The act is drama of drama. Turning point occurred in this scene of this drama's emotional. First paragraph is Christina's solo part, flirtatious and vibrant. The second part is the chorus of Phantom and Chritina, representing the romantic and ambigous love between them. In the last part, the theme of Phantom of the opera sounded, Phamtom taked off the mask, climax of the opera. We decided to use red and blue to represent actor and actress. Combined with site characteristics of our group, we want to show actor and actress' respectively emotional change by controlling the light.

作者 Designers

张佳择　高子晗　倪凌　周逸昀　李英子　张倩　唐恺鑫　彭心悦
朱冰洁　任宇恒　郭晧阳　胡艺龄　李宛蓉　周雨桐　段睿妍　翁子健

分镜头脚本设计 Storyboarding design

第八组 The point of no return

一共三排灯 3*8 灯
A 组 第一排
B 组 第二排
C 组 第三排

0:00–0:53
Past the point of no return(4)
C 红色
最后一排灯 两盏灯 从两端亮到中间

no going back now(4)
B 红色

our passion play has now at last begun(8)
边缘到中间随机亮哪一排
最后两拍和在一起不变 延长

(dengdengdeng)
(dengdengdeng)
每一个 deng 灭一部分灯，到最后一个后全灭

Past all thought of right and wrong(4)

one final question:
how long should we two wait before we're one?(8)
重复第一小节手法

When will the blood begin to race the sleep in gbudburstin to bloom?
When will the flames at last consumeus?

(dengdengdeng)
(dengdengdeng)
重复第一段的手法

0:53–1:22
BOTH:
Past the point of no return(4)
一盏盏亮，区分红蓝色
the final threshold(4)

the bridge is crossed
so standand watch it burn
（一盏盏 全亮布满整个舞台）
We've passed the point of no return
（慢慢向两边的两盏灯熄灭，一盏红色，一盏蓝色）

1:22–1:32
(dengdengdeng)
(dengdengdeng)
这两盏灯晃动起来

1:32–1:45
PHANTOM:
Say you'll share with me
one love, one life time
（代表男主的蓝色区域的灯开始扩散）
Lead me, save me from my solitude
（代表女主的红色也开始扩散到男主角的区域，此时舞台上有红蓝两种颜色的灯，但是混合在舞台上）
Say you want me with you
here beside you
（红蓝灯中的另一种颜色也逐次打开，形成每一盏灯都有红蓝亮色的感觉）

Anywhere you go let me go too
一列一列或者一排排灭向下面灭掉所有的灯
Christine, that's all I ask of...

02:15–2:36
主旋律，开始把幕布给拉起来

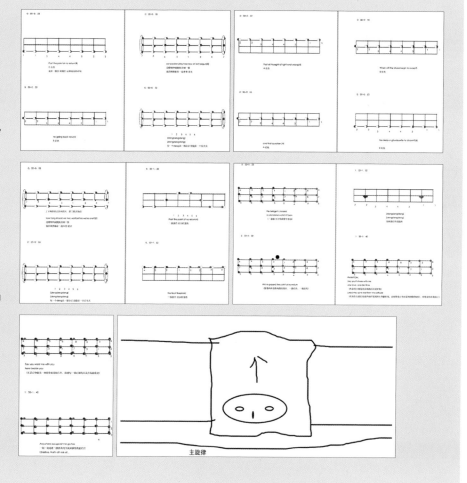

设计草图 Sketches

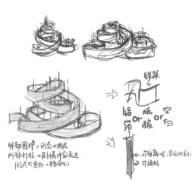

材料构成 Material

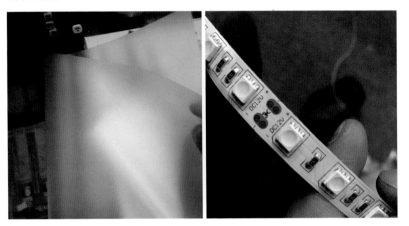

制作过程 Construction

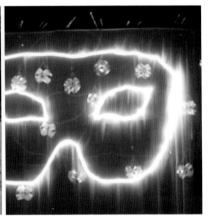

现场安装 Installation

分镜头演出效果
Storyboarding

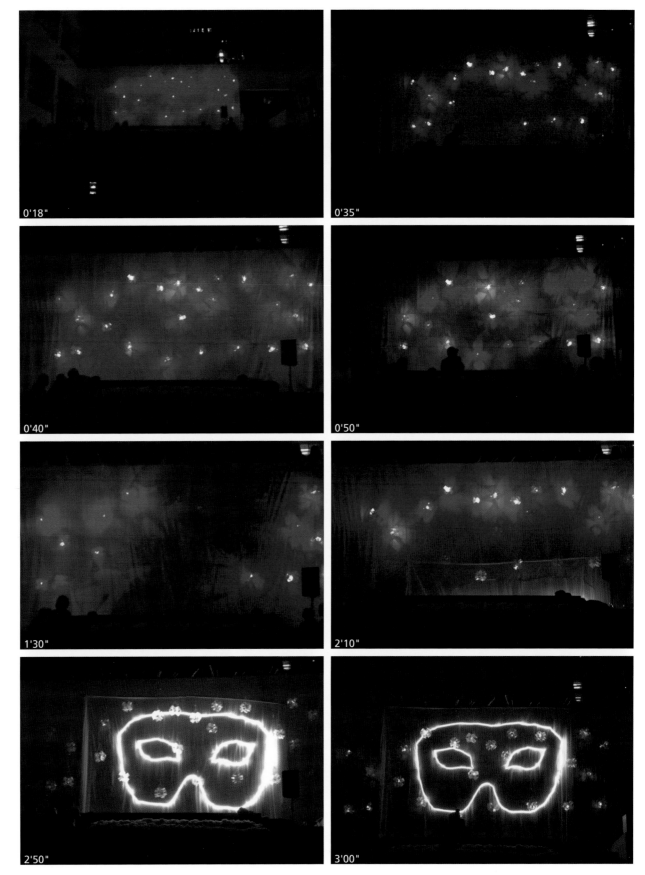

第九幕
仲夏夜之星 Finale

Finale 的乐曲以八音盒为开端，清脆空灵，跳跃感强，犹如仲夏之夜一片苍穹之下的点点繁星。紧接着一段悠扬的弦乐渐渐响起，宛如一缕缕皎洁的月光洒在云端，缠绵连续，与忽明忽暗、跳跃感强的繁星交相呼应。音乐的高潮部分由原来的清脆婉转突变得温暖激荡，于是日出、星陨，所有云朵散发出耀眼的金色，直至音乐缓缓消失。
我们组最终以硫酸纸折出星星的轮廓，将草帽 LED 灯置入其中，再将 LED 颗粒放入摆在平台矮墙上的棉花中，配上 LED 灯带营造出云朵意向。而男主角和女主角戴着面具与皇冠，两者互动，情感含蓄流露。

The music starts with a piece of clear and lively music box, which presents a picture of a starry summer night. The piece segues into melodious strings, just like moonlight shining through the clouds and works in concert with the stars. Then the climax turns to be stirring and thus we present a picture of a rise of the sun, a fall of stars and a glare of clouds.
Our team used acid paper to create the shape of stars. Cotton and LEDs are used to create clouds. And the two leading actors were represented by a mask and crown.

作者 Designers

金世煜　王兆一　陈路平　冯　田　高雨辰　王旭东　何斌贤　聂方达
谢成龙　庄铭予　刘雨婷　唐奕诚　陈　俐　张万霖　郝明宇　卡那库亚

分镜头脚本设计 Storyboarding design

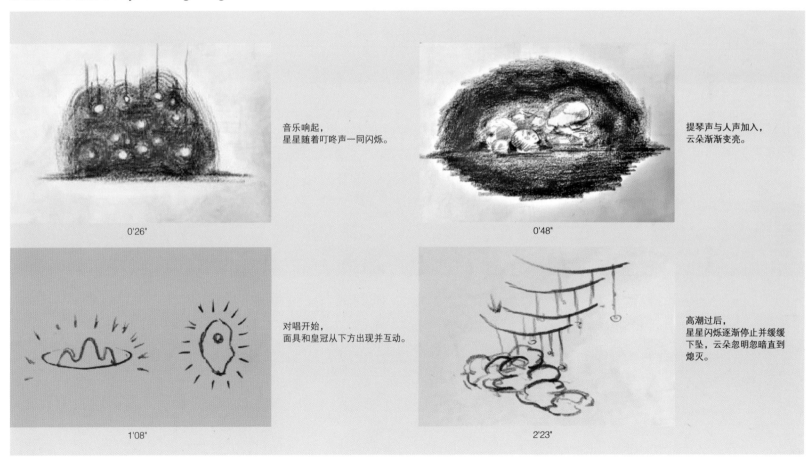

0'26"　音乐响起，星星随着叮咚声一同闪烁。

0'48"　提琴声与人声加入，云朵渐渐变亮。

1'08"　对唱开始，面具和皇冠从下方出现并互动。

2'23"　高潮过后，星星闪烁逐渐停止并缓缓下坠，云朵忽明忽暗直到熄灭。

设计草图 Sketches

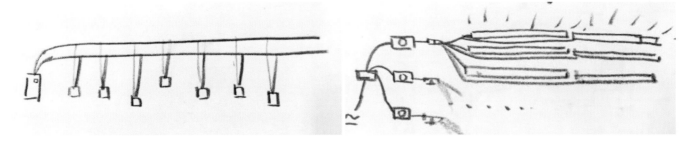

材料构成 Material

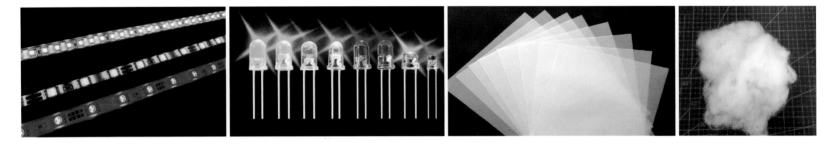

制作过程 Construction

现场安装 Installation

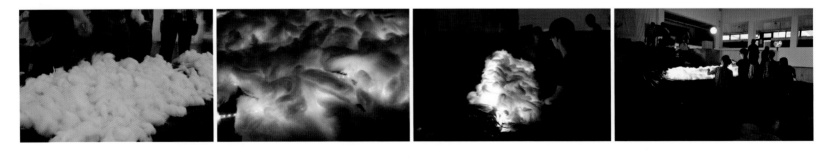

分镜头演出效果
Storyboarding

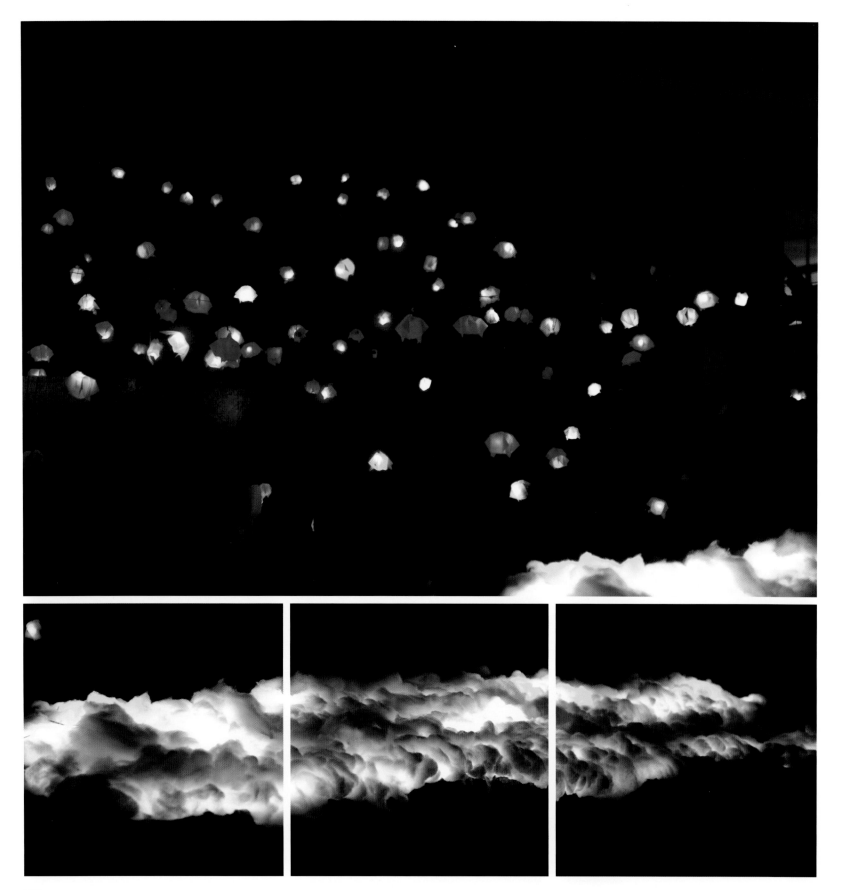

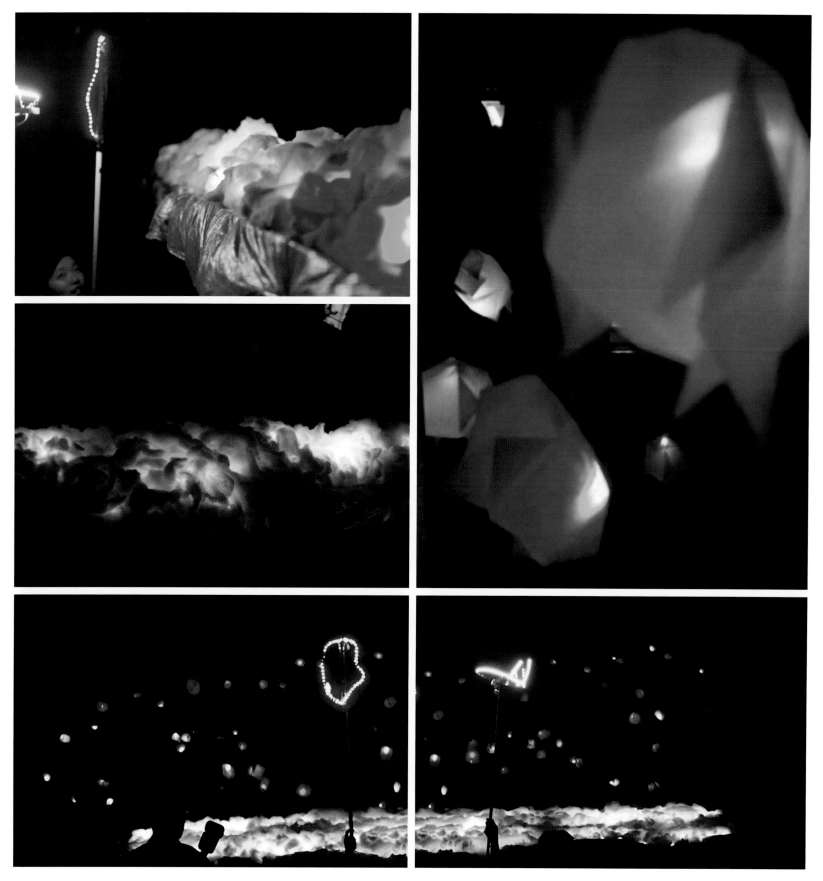

第十幕
化妆舞会 Masquerade

在 Masquerade 一幕中，我们想用光的语言去再现这一盛大的场景，通过光与同学们舞蹈动作的配合给人以感染力。面具、妆容、扇子、衣服的配合再加上整场演出的气氛渲染，让这一幕给人热闹非凡、庄重而欢快的感受。

In the scene Masquerade, we want to recreate this great ball by using the language of light. The combination of light and choreography touches people deeply. The masks, fans, gowns together with the gowns and makeup gave the audience the impression of excitement, solemnity and cheerfulness.

作者 Designers

分镜头脚本设计 Storyboarding design

0'05"
灯光亮起,女舞者搀扶男伴依次入场,成为全场焦点。
0'25"
舞者们第一次入场的集体群舞,严肃庄重,展示了假面舞会的高贵隆重。

0'30"
第一次群舞后的集体亮相,宣告正式严肃的群舞结束,预示气氛将开始变得活泼随意。
1'00"
舞者们开始群舞时,身上的光为全场最亮最瞩目,吸引所有人的注意。

1'15"
舞者们走到观众中央,起舞,全场灯光亮起。
舞者们又重新回到开场时群舞的舞台,重又严肃庄重,将全场气氛推至最高潮。

3'00"
全场气氛达到最高潮,观众和舞者们一同起舞。
4'00"
舞会结束,意犹未尽的舞者和观众们一同开始跳起圆舞曲。

设计草图 Sketches

材料构成 Material

制作过程 Construction

现场安装 Installation

分镜头演出效果
Storyboarding

谢幕与颁奖 Curtain calls and awards

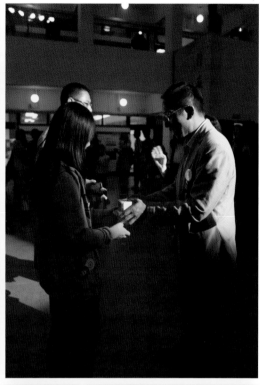

历年教学回顾 Review of years of teaching

2002

课程要求学生对生活中的光环境进行调研，利用实验室的设备，采用不同的材料和照明方式，截取有意思的片段进行表现。这是本课程体系首次尝试脱离常规的教学模式，让学生亲自体验光照场景，理解物理量的视觉意义，以培养他们运用所学知识和设计技法解决实际问题的能力。

2003-2005

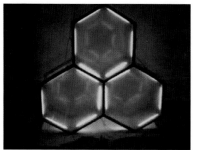

光与水立方——北京奥运会国家游泳中心（水立方）室内及立面光环境设计

西安汉阳陵遗址保护展示厅及周边环境照明设计

在建筑光环境实验性教学体系探索初期，光艺术装置教学是照明专门化毕业设计中的重要环节。学生们根据自己的方案制作整个建筑或局部片段的缩尺灯光工作模型，在建立光环境概念的同时，掌握第一手光度数据资料，更好地将光度物理量与实际效果相联系。

2006-2007

利用传统光源如白炽灯、卤素灯、荧光灯等，易于市场选购的光源和实验室的设备，将光作为生动的设计语汇进行创作。其形态、光影、明暗构成动态或静态的光照图示。学生最终将成果制作成幻灯片或多媒体动画，展示他们对光与材质、空间的认识和理解。

2008

由于照明技术的发展，光艺术装置实验环节引入低碳环保的LED光源作为创作素材。教学场地的LED显示屏成为同学们的创作基础，每组学生分配25cm×25cm大小的显示屏，他们在充分了解LED发光特点、材料特性的基础上进行光艺术媒体界面的创作。

| 2009 | 2010 | 2011-2012 | 2013-2015 |

2010世博会前期的学生作业,在世博文化中心两个入口大厅背景墙得到深化应用,最终作品完全由师生现场制作完成,获得了应用单位的高度认可,取得了良好的社会反响。并在若干年后被应用于实验室人居空间健康照明的研究课题中。

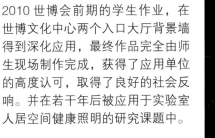

"声光SHOW"大型装置中,学生们在了解语音输入、声音信号识别、人机交互等原理的基础上,自己编排声光效果,再利用矿泉水桶作为发光像素点,上演了一场别开生面的"声光秀"。

课程更加注重实践性,学生亲自动手焊接LED颗粒并连接电源。装置作业的规模、成熟度及艺术效果得到了加强。并在最后加入评审展示的环节,评审老师根据展示现场效果给出课程成绩。

从2013年开始装置作业有了主题,学生们在限定主题、场地、时间、装置制作材料、LED颗粒颜色和数量的条件下,以小组为单位,制作光艺术装置,结合"徒手控制"的动态编排,探索光影如何与空间、环境、音乐互动,探讨光与色彩如何改善生活环境、诠释传统文化、演绎戏剧情节。发现光的多维度应用,体验光的无限魅力。

图书在版编目（CIP）数据

2015·流光魅影 / 郝洛西等著. -- 上海：同济大学出版社，2016.6
（同济大学建筑学专业"建筑物理（光环境）"教学成果专辑）
ISBN 978-7-5608-6200-2

Ⅰ.①2… Ⅱ.①郝… Ⅲ.①建筑物理学—教学研究—高等学校—文集
②建筑光学—教学研究—高等学校—文集 Ⅳ.①TU11-53

中国版本图书馆CIP数据核字(2016)第025391号

同济大学建筑学专业"建筑物理（光环境）"教学成果专辑
2015·流光魅影

郝洛西 等 著

责任编辑	张　睿
责任校对	张德胜
装帧设计	李　丽
APP制作	今尚数字
出版发行	同济大学出版社（www.tongjipress.com.cn）
地　　址	上海四平路1239号（200092）
电　　话	021-65985622
经　　销	全国各地新华书店
印　　刷	上海丽佳制版印刷有限公司
开　　本	787mm×1092mm　1/12
印　　张	37.33
字　　数	940000
版　　次	2016年6月第1版
印　　次	2016年6月第1次印刷
书　　号	ISBN 978-7-5608-6200-2
定　　价	210.00元（全三册）

向 2015 国际光年致敬

同济大学建筑学专业"建筑物理（光环境）"
教学成果专辑

2014·光影时节
LIGHT SHADOW AND SEASON

郝洛西 等 著

同济大学出版社
TONGJI UNIVERSITY PRESS

目录 Contents

Part 1	Prologue 写在前面		6
Part 2	Task introduction 作业简介		22
Part 3	Students' work show 学生作品展示		24
	01	Summer with ink and wash 立夏・水墨意夏	26
	02	Imperfect perfection 小满・未满	32
	03	Dew is busy with people 芒种・露随人忙芒	38
	04	Summer Solstice spinning cirrus and flying light 夏至・蔓转流光	44
	05	Life like summer flowers 夏至・生如夏花	50
	06	Shadows' wander with wind 夏至・影踱风舞	56
	07	Lotus in the rain 夏至・风曳雨荷	60
	08	Twist for the whole summer 夏至・转一夏	64
	09	Summer tree 夏至・夏树	68
	10	Summer daydream 小暑・夏日梦舞	72
	11	Hot sun breaking rocks 大暑・炽日裂岩	78
	12	The falling of one leaf heralds the autumn 立秋・一叶知秋	84
	13	Waving in bamboo ocean 处暑・泛波竹海	90
	14	White Dew, the whole ground 白露・白露满地	94

15	Waltz of winged miration 秋分·雁来燕去	98
16	Dew cream dream 寒露·白露为霜	102
17	Blue blue glass moon 霜降·落幕飞霜	108
18	Winter Begins 立冬	112
19	Light Snow 小雪	116
20	Great Snow 大雪	122
21	The Winter Solstice 冬至	126
22	Song of moon & ice 小寒·浮月寒冰	132
23	Great Cold 大寒	136
24	Ice melts into serpentine water 立春·解冰曲水	140
25	The rain lyric 雨水·雨的抒情	144
26	The waking of insects 惊蛰	148
27	The Spring Equinox yin-yang fish dance 春分·春春鱼动	152
28	Cloud and rain in Qingming 清明·蔚云细雨	156
29	Flourishing season 谷雨·生长时节	160
Part 4	Teaching highlights 教学花絮	164
Part 5	Show awards 展示颁奖	166
Part 6	Review of years of teaching 历年教学回顾	168

Light and shadow, heart and shape

Ever since the existence of light, numerous shapes of shadows follow.
Ever since the existence of architecture, light and shadows accompany it.
As the Chinese sayings go: "Hou Yi shooting the suns", "Li Bai staring at the moon" – natural light gives buildings vivid lives; "Ying people holding the candle", "Watching lanterns on the Lantern Festival" – artificial light expands the horizon of cities and architecture. In the industrial society, "Edisons" lit up the night sky and every corner of buildings with electrical lights; while in the information society, new technologies emerge. LEDs, with less energy consumption and more flexibility, emit even more beautiful light and colors, leaving people soul-stirring light and shadow.
To help future architects understand the charm of light and shadow, Professor HAO Luoxi and her team have been teaching the course "Architecture Physics: Lighting Environment" for 15 years. This is a fundamental, 17-hour course for sophomores of College of Architecture and Urban Planning (CAUP), Tongji University. The course hour is short, however, both teachers and students exhibited great involvements, and very much enjoyed the wonderful teaching and studying processes. The three books presented here in front of us today, including "2015·Let's be Your Light", "2014·Light Shadow and Season" and "2013·The Light of The Antarctic", are the achievements from cooperation between teachers and students during the years 2013 – 2015.
We are moved by such achievements. These are teaching and experiments with rich characteristics of the time. They are combinations of art and technology, through cooperation between teachers and students.
They think and work: arrange light sources, design devices; they learn from history and raise questions to future, wandering around in the free realm of light and shadow, being intoxicated. They organize exhibition visits, design dramas, prepare music, and turn pure lighting technology into comprehensive art exhibitions as well as lighting festivals for teachers and students.
All of these stemmed from their desire to pursuit technology improvements, and their wish mood to seek various art forms. The forms to express vision and longing comes from their desire and wish.
We look forward to seeing Professor HAO and her colleagues to continue the innovations and "emit more gorgeous light". We appreciate people who offered their help during the series of teaching processes. We remember the warm shadows they left, and wish students great achievements in future study and creation, and "exhibit more beautiful colors".

Prof. Dr. LI Zhenyu
February 22, 2016 (The Lantern Festival)

光与影，心与形

自从有了光，便有了千姿百态的影。

自从有了建筑，光影便一直与之相伴。

后羿射日，李白对月，自然光赋予建筑有声有色的生命；郢人秉烛，元宵看灯，人工光拓展了城市和建筑的时空。进入工业社会，爱迪生们用电灯照亮了夜空，照亮了建筑的每一个角落；进入信息社会，新技术层出不穷，LED 以更少的能源，更便捷的组合，发出更加美妙的光色，留下更加动人的丽影。

为了让未来的建筑师们理解光影的魅力，郝洛西教授与她的团队开设"建筑物理（光环境）"已经有十五年了。这是一个 17 课时、面向同济大学建筑系本科生二年级的专业基础课程。课时虽少，但参加课程的老师、同学们却非常投入，非常享受这样美好的教学过程。今天呈现在大家面前的三本书《2015·流光魅影》《2014·光影时节》《2013·南极之光》，就是 2013–2015 年期间师生共同合作的成果。

这样的成果，让我们为之感动。这是富有时代特征的教学和实验，是艺术和技术的结合，是老师和学生的合作。

他们动脑动手，布置光源，设计装置，向历史学习，向未来提问，徜徉在光和影的自由王国里，如痴如醉；他们组织展览，策划剧情，配制音乐，把单纯的光的技术，变成了综合的艺术展示，变成了师生共同的光影节日。

这一切，都是因为他们有追求技术进取的心愿，追求艺术多态的心境。有了心，才有了表达情怀的形。

我们期待郝老师和她的同事们不断创新，发出更加绚烂的光；我们感谢给予这一系列教学活动各种帮助的人们，记住了大家留下的热情的影；我们祝愿同学们在未来的学习和创作中大有作为，呈现出更加美丽的光与色。

院长
2016 年 2 月 22 日元宵节改定

Inheritance and development

The book series on your hands belongs to the series of Tongji "Architecture Physics: Light Environment", which is the outcome of fifteen years' teaching exploration and innovation. This book, for the first time, attempts to introduce the format of digital media besides the paper-based publication. Readers can download the APP and have an intuitive experience of the review site, and fully appreciate the splendor of each work in the books. This series includes three books: the first book is "2015·Let's be Your Light" published in 2015; the second book is "2014·Light Shadow and Season" in 2014; the third one is "2013·The Light of The Antarctic" in 2013. Due to the limited space, teaching achievements before 2012 are not included in this publication; however, years of work still exudes the unique light, such as the large-scale installations"sound and light show" during the 2010 Shanghai World Expo; the work "bloom", after the demonstration in Expo cultural center, continues its applications in healthcare buildings, passing the infinite charm of light and space.

As an important fundamental course, "Architecture Physics: Light Environment" is designed for the sophomores with an architecture major. Given only 17 credit hours, the course is taught in three main aspects: (1) light, color and visual environment; (2) the light source and lamps selection (laboratory teaching); (3) quantity and quality of lighting. As there are courses "Interior Lighting Art" and "Daylight and Architecture" for senior students, this course can be seen as the entry-level courses for students to have a feeling of light and illumination. Over the years, the coursework topics are carefully designed by the teaching team, focusing on the light and optical properties of materials, shape, light-emitting device, the control concept, light drawings and expression, and other rich and comprehensive understanding. This helps students to have innovative thinking, practical hands-on operation, and the attitude of pursuing excellence in design works. As a team, students cooperate to complete the job. Everything is designed for students' self-learning, exploration, and practice.

Although the course is limited to six weeks and three credit hours per week, the students are sophomores who just stepped into the design world, and furthermore, the students must ensure that their main courses are not much affected, the students succeeded to finish the good work in the form of team work, with the limitation of defined theme, venue, time, number and color of LEDs, the conditions of environmental protection waste materials. The last day of teaching is for the show and assessment of homework. Therefore it can be treated as a time-limited quick architecture design, all the on-site dynamic changes are students' "hand control". The huge number of constraints are indeed challenges to both teaching and learning sides, and this is why I, as the teacher for this course, always be amazed by students' works.

We appreciate the support from the president of the Tongji University Press, Prof. ZHI Wenjun. He always pay close attention to the team's teaching efforts. Prof. ZHI have suggested us to publish our teaching efforts many times. Due to the fact that it is really difficult to reflect the content and expression of students' coursework by paper-based publications, this effort had been delayed to seeking the opportunity of digital-media publication. This summer, Prof. ZHI and the president of Jin Shang Digital, Mr. Xiaobo Xing, had a discussion with me at the Tongji University Press regarding the publication of this teaching effort. Ultimately, we decided to utilize both paper-based and digital-based formats to publish the three books, to share our experience and results from our 15 years of teaching on the course "Architecture Physics: Light Environment" .

Special thanks to CAUP colleges, graduate and undergraduate students, and friends from industry who participated in this teaching work. Especial thanks to LIN Yi and CUI Zhe, for their contribution in curriculum construction.

As 2015 is the "International Year of Light" declared by the United Nations, to commemorate "the father of optics" - Ibn Hazm, who wrote an important optics book a thousand years ago, and a series of important finding in optics history, we would like to take the opportunity to pay tribute to the "International Year of Light"!

HAO Luoxi
12/15/2015

传承与开拓

您手中的这套书是同济大学"建筑物理(光环境)"这门课程 15 年教学探索与创新实践的成果汇编。这套书尝试在纸媒出版的基础上引入数媒,读者下载 APP 就可直观体验到作品评审时的现场情况,充分领略每个作品的光彩。本套书共计三册,第一册是 2015 年的作业汇编《2015·流光魅影》;第二册是 2014 年的作业汇编《2014·光影时节》;第三册是 2013 年的作业汇编《2013·南极之光》。由于出版篇幅所限,2012 年之前的教学成果没有列入此次的出版计划,但那些年的作品依然散发着特有的光芒,如 2010 年上海世博会期间的作业"声光 SHOW"大型装置;再如当年的作品"绽放",历经世博文化中心的演绎应用,作为后世博的再应用,继续在医疗健康建筑中,传递着光与空间的无限魅力。

作为建筑学一门重要的专业基础课程,"建筑物理(光环境)"面对建筑学专业二年级的本科生授课,鉴于只有 17 个学时,主要讲授三个内容:光、颜色与视觉环境;光源与灯具选型(实验室授课);照明的数量与质量。对四年级本科生开设有"室内照明艺术"和"日光与建筑",因此可以看作是同学们接触光与照明知识的入门课程。历年课程作业选题来自教学团队的精心设计,重点在光、材质及光学特性、形态、发光器件、控制概念、光照图式及表达等丰富而综合的理解,培养学生开拓创新的思维能力、实践操作的动手能力、追求卓越的设计态度。作业完成以小组为单位,培养学生协同攻关的能力。一切向着自主学习型、设计探索型和实践应用型的学习方式迈进。

尽管只有为期六周、每周三个学时的授课,尽管他们只是初涉设计的二年级学生,但他们在保证专业课程设计作业不受影响的情况下,以团队合作的形式,在限定主题、场地、时间、LED 颗粒数量和颜色、环保废弃材料的条件下,完成最后的作业。教学的最后一天进行作业展示及评审。因此也可以说,这是一个类似建筑设计快题的限时作业,现场所有的动态变化均是学生们的"徒手控制"。如此多的限定条件,对教与学来说都是莫大的挑战,我作为任课教师和大家一样,对学生们最后呈现的作品总是每每发出由衷赞叹。

承蒙时任同济大学出版社社长的支文军教授一直以来对团队教学的关注,多次讨论将该教学的成果汇编成册出版,但苦于仅凭纸质的确很难全面反映该课程学生作品的内容与表达,于是就心有不甘地一直寻求数媒出版的可能性。今年暑假期间,支教授连同"今尚数字"邢小波先生约我到同济大学出版社,一起探讨出版事宜。最终我们决定以纸媒加数媒的出版形式,由张睿老师担任责任编辑,出版这三册书,分享 15 年来承担的"建筑物理(光环境)"课程的教学成果。

在此特别致谢团队中共同进行课程建设的林怡、崔哲两位老师,还有参与教学工作的学院老师、团队研究生、历届本科生及企业界朋友。

今年适逢联合国宣布 2015 年为"光和光基技术国际年(简称国际光年)",以纪念"光学之父"伊本·海赛姆的光学著作诞生一千年及光科学历史上一系列重要发现,借这套书的出版,向国际光年致敬!

任课教师
2015 年 12 月 15 日

小滿

桑葉正青肥，
蠶兒篋籠圍。
石榴開口笑，
苦菜遍山陲。

立秋

宿草盈珠露,
青槐一叶飘。
凉风吹瑟瑟,
知了唱秋高。

芒種

退步插秧田,
螳螂壟上閒。
農人忙此季,
新麥早開鐮。

夏至

垂柳映荷塘,
風吹菡萏裳。
竹簾涼月色,
蛙叫夢長長。

小暑
山風搖翠竹,
濕氣結云嵐。
倏忽及時雨,
炎消聞夜蟬。

大暑

水艾含香味,青蒲吐劍云。
年年唯六月,大雨落傾盆。

谷雨

谷雨采新茶,
村姑日不暇。
朝来坡上过,
桑葚已红颜。

立夏

牡丹花日零,
紫燕語中亭。
午后雷聲動,
絲絲雨霧蒙。

小雪

秋草牧羊肥，
絨毛厚厚堆。
雖逢入冬雪，
何懼冷相衰。

大雪

昼短夜何長，
冬眠万物藏。
朝來天降雪，
四野白茫茫。

冬至

冬至一阳回,
當催游子歸。
下元誰祭祀,
墳上紙灰飛。

小寒

歲歲念梅來，
今冬如季開。
聞香欲尋去，
摘句入詩栽。

作业简介 Task introduction

一、作业题目

光影时节——致敬 2015 国际光年

二、作业目的

通过对中国传统二十四节气的学习和了解，动手制作光艺术装置，以契合二十四节气的特点与氛围，传播中华文化，致敬 2015 国际光年。

作为低碳环保的新光源，LED 是这次作业所采用的发光器件，结合生活中常见的材料，基于 LED 发光特点和材料的光学特性，尝试光、材质及形态的创新设计。在这个实验性教学过程中，发挥团队精神，努力探索，通过焊接 LED 的操作实践以及光装置的创意设计，观察各种材料、光线及环境的互动关系，发挥大胆和丰富的想象，用光描绘出一幅幅绚烂的艺术构成，呈现出一场独特的视觉盛宴。

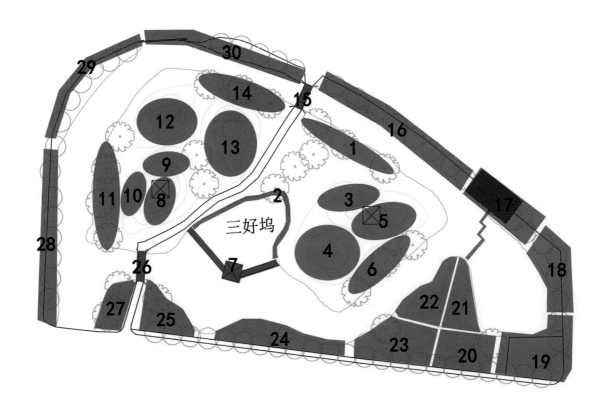

三、作业内容与方法

1. 使用材料

材料不限。推荐使用生活过程中产生的废物，如啤酒瓶、易拉罐、矿泉水瓶、方便面袋子、垃圾袋等，循环利用，力求对环境的影响最小化。

2. 光源选择

LED 芯片，共有 10 种不同光色，每组数量统一，每种光色各 10 颗，共 100 颗。

3. 方法

LED 焊接：利用电烙铁等工具将 LED 芯片与电线焊接。

装置设计：探索光与材料的特性，进行光艺术效果实验，选择符合主题的效果。

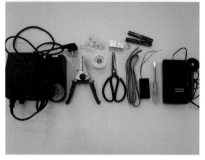

4. 要求

以"光影时节——致敬 2015 国际光年"为主题，将中国传统二十四节气的特点、寓意、气氛融入光艺术装置中，在三好坞的园林环境中进行展示，要求与基地环境景观要素高度结合。注意可运输性、可拆卸性，也可采用 DIY 方式。

四、作业组织

以班为单位分成工作小组，每组 5-6 名学生。

五、作业成绩评定

按照工作小组进行公开评审，每小组一个成绩。

六、作品展示

展示时间：2014 年 6 月 21 号（周六）19:00-22:00

展示地点：同济大学三好坞

评审颁奖：文远楼 215 室 21:00-22:00

技术支持：上海亚明照明有限公司
　　　　　上海科锐光电发展有限公司

学生作品展示 Students' work show

作业主题

以"光影时节"为主题,将中国传统24节气的特点、寓意、气氛融入到光艺术装置中,在同济大学三好坞园林环境里进行展示。作业装置基于LED发光特点和材料的光学特性,考虑基地环境景观要素,尝试光、材质及形态的创新设计,突出低碳环保的理念。在实验性教学过程中,鼓励学生发挥团队精神,努力探索,通过创意设计及操作实践,用光描绘绚烂的艺术构成,呈现独特的视觉盛宴。

任课教师 郝洛西 教授
技术支持 上海亚明照明有限公司
上海科锐光电发展有限公司

地点 同济大学三好坞
时间 六月二十一日(周六) 晚七时至十时

光影时节
致敬2015国际光年
2012级建筑学专业
建筑物理光环境
光影构成作业评审展示

作业主题

以"光影时节"为主题,将中国传统24节气的特点、寓意、气氛融入到光艺术装置中,在同济大学三好坞园林环境里进行展示。作业装置基于LED发光特点和材料的光学特性,考虑基地环境景观要素,尝试光、材质及形态的创新设计,突出低碳环保的理念。在实验性教学过程中,鼓励学生发挥团队精神,努力探索,通过创意设计及操作实践,用光描绘绚烂的艺术构成,呈现独特的视觉盛宴。

任课教师 郝洛西 教授
技术支持 上海亚明照明有限公司
上海科锐光电发展有限公司

地点 同济大学三好坞
时间 六月二十一日(周六) 晚七时至十时

光影时节
致敬2015国际光年
2012级建筑学专业
建筑物理光环境
光影构成作业评审展示

光影时节

致敬2015国际光年
2012级建筑学专业
建筑物理光环境
光影构成作业评审展示

时间 六月二十一日（周六）
　　　晚七时至十时

地点 同济大学三好坞

作业主题

以"光影时节"为主题，将中国传统24节气的特点、寓意、气氛融入到光艺术装置中，在同济大学三好坞园林环境里进行展示。作业装置基于LED发光特点和材料的光学特性，考虑基地环境景观要素，尝试光、材质及形态的创新设计，突出低碳环保的理念。在实验性教学过程中，鼓励学生发挥团队精神，努力探索，通过创意设计及操作实践，用光描绘绚烂的艺术构成，呈现独特的视觉盛宴。

任课教师 郝洛西 教授
技术支持 上海亚明照明有限公司
　　　　　　上海科锐光电发展有限公司

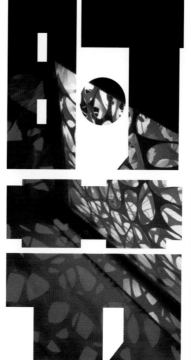

立夏・水墨意夏
Summer with ink and wash

作者 Designers

徐洲　钱蕴安　何昱婧　曾鹏程　王梅洁　曾鹏飞

概念原理 Concept

我们在夜色中漫步，在虫鸣中捕捉灵感，月下斑驳的枝桠的魅影随清风晃动，极具仲夏夜摄人心魄的美。紧抓这稍纵即逝的灵感，结合基地茂盛的竹林，我们以宣纸为面，借光影为笔，一同创作了迷蒙氤氲的水墨意夏。希望中国水墨画独有的宁静淡泊，和着昏暗迷魅的灯光，在炎炎夏日中给纳凉的过客带来美和清凉。

We walked in darkness, looked for inspiration in the insect chirp. Mottled shadow of the branches swaied with wind in the moonlight. It was so amazingly beautiful and we were touched deeply. So with the lush bamboo forest around the site, we used light and shadow to draw a ink and wash painting. We hope this Chinese classical atmosphere can bring people a feeling of beauty and cool in the muggy summer night.

材料构成 Material

设计草图 Sketches

制作过程 Construction

成果表现 Realization

小满・未满
Imperfect perfection

作者 Designers

李颖劼　张家宇　洪邦建　陈思宇　陈正阳　斯恩兴

概念原理 Concept

"小满"意为果实青涩初现，尚未成熟完满的状态，也是其在节气的主要特征。"小满未满"既代表着一种趋向成熟的激情，也有含苞等待的含蓄暧昧，这成为我们光装置的设计目标。同时，考量基地的环境特征，结合"岩石"、"树木"以及"湖"，我们选择镂空装置对LED的光芒进行转化，以岩石、地面和树丛作为光线的承接面，从而与基地充分结合并创造丰富的光影效果。在水面上，利用半透明的拷贝纸处理装置，达到"水路同辉"。LED依托镂空的光装置，避免光线直射，在"遮掩"与"开放"之间形成虚实对比，感受"将满"的期待与"未满"的暧昧完美结合。

Grain buds means the seeds of fall grain are becoming full but not ripe yet which is also the main feature of this solar term. This "becoming" full but not ripe represents not only a trend of mature, but also a kind of implicite motion of waiting. And this becomes our goal to design lighting devices. Meanwhile, in consideration of environmental characteristics of our place, as well as rocks, trees and water, we choose hollow-carved devices to display the light of LED. To totally combine with our place and create a gorgeous scene with light and shadow, rocks land and trees are choosen as the receiving surface for light. At the same time, with the assistance of translucent copy tissue, the device on the water successfully achieved the fasciante effect of co-enlightenment on both water and shore. Relying on the hollow-carved equipment, light rays casted by the LEDs are partially blocked, and concequencely set the comparison between outgoingness and modesty. Thus we sense the ambigiousness of regret and the great expectation of content.

材料构成 Material

设计草图 Sketches

 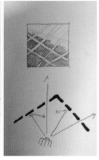 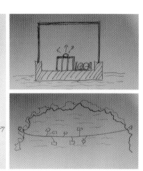

制作过程 Construction

 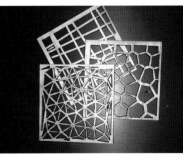  

成果表现 Realization

芒种·露随人忙芒
Dew is busy with people

作者 Designers

黄炜乐　朱旭栋　刘育黎　何星宇　李振榮　葛梦婷

概念原理 Concept

芒种最大的季候特点是梅雨，所以我们借用悬挂的细管来模拟"雨"。在灯光的设计上，我们希望呈现芒种的过程态，芒种含义是收获然后接着播种，有一种紧张的气氛，所以顶部和底部运用暖色象征已经收获的和还未收获的庄稼；中部冷色的灯光象征即将洒下的种子，因为还在空中所以能很好地表现过程态；最底下的红光代表红红火火的意味，表现出人们忙碌的气氛。在基地的考虑上，我们希望借三个岔道的汇聚让装置在平台上达到一个高潮。

Grain is characterized by rain, so we borrowed rain characteristics in the form of tube to simulate climatic characteristics; in lighting design, we hope to show Budding process state. Budding means is harvested and then planted. There is a tense atmosphere, therefore, the use of warm colors symbolizes the top and bottom has not yet been harvested and harvest crops respectively. The central symbol of the upcoming cold light shed seed, because it is still in the air can be a very good performance process state, the most representative red bottom booming fire mean, showing people busy atmosphere. On the consideration of the base, we hope that the convergence of the three forks to allow the device to reach a climax on the platform.

材料构成 Material

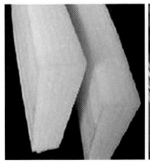 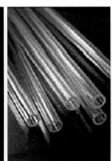 

设计草图 Sketches

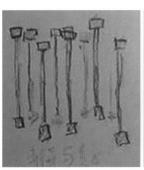  

制作过程 Construction

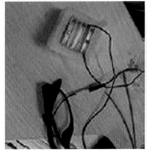    

成果表现 Realization

夏至 · 蔓转流光
Summer Solstice
spinning cirrus and flying light

作者 Designers

林昱宏　桑铖卓　徐泽炜　孙逸群　张锋　饶鉴

概念原理 Concept

"至者，极也"，太阳极盛，万物生长，取"藤蔓"为意象，虚幻更替，盘旋生长。制作方法是取废旧光盘碎片，插于泡沫上，放入"藤蔓"内部，制作成旋转装置，利用光盘的反射纹理及色光的叠加在拷贝纸上映射出来，模拟轮回更替。夏至，最繁盛之时，便也是开始衰减之时，盛极必衰，循环往复，引人无限遐思。

"Solstice, is extreme", the sun reaches the top, all things grow, we take "vine" as the image, which has unreal changes, spiral growth. Manufacture method: take the waste disk fragmentation; insert it in the foam; put it into "vine"; make them into a rotating device. Reflection texture and color of the disc is superimposed on the copy paper mapped out, simulating cycle change. The summer solstice, is the time which is the most prosperous, and also the starting point of the attenuation. Moving in circles, fetching the infinite reverie.

材料构成 Material

设计草图 Sketches

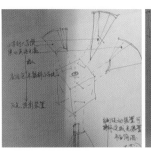 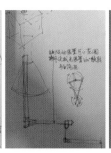 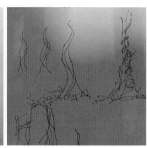

制作过程 Construction

    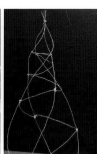  

成果表现 Realization

夏至・生如夏花
Life like summer flowers

作者 Designers

张 克　　程 锦　　郑兆华　　吴鹏飞　　马一茗　　沈逸飞

概念原理 Concept

我们这次的光装置的主题节气是夏至。在每年公历6月21日或22日这天，太阳直射地面的位置到达一年的最北端，几乎直射北回归线，此时，北半球的白昼达最长，且越往北越长。此日日影最短，所以我们决定选取太阳这一意象，制作出"太阳花"的效果。并有感于夏至日的暴雨、梅雨天气、高温桑拿等气候特点，以"生如夏花"这个主题来体现它的活力所在。至于实现方式，我们借用羽毛球的空隙在灯光下投射出的影子来表现"太阳花"的效果。整个装置用细线悬挂，人流经过时便可与之发生互动，形成疏影摇曳的唯美景象。

Our theme is the summer solstice. Every year on June 21 or 22, the sun casts the extreme north of the land. In the northern hemisphere the day reaches the longest, and the longer the farther north. This day, shadow is the shortest, so we decided to choose the sun as a image, produce the effect of "sunflower". Being aware of the summer solstice of torrential rain, the rainy weather, climate characteristics, we choose "life like summer flowers" as the theme. We use badminton gaps to cast a shadow to represent the effect of "sunflower". The whole device with hanging fine line can interact with the occurrence of stream of people passed, form a thin film of swaying beautiful sight.

材料构成 Material

设计草图 Sketches

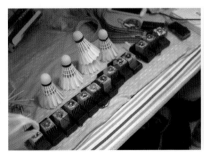 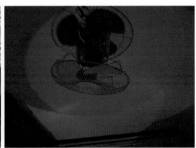

制作过程 Construction

成果表现 Realization

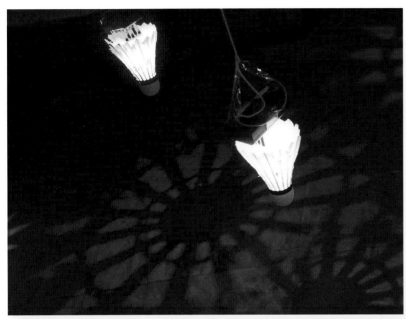

夏至·影踱风舞
Shadows' wander with wind

作者 Designers

李霁欣　徐蒙恩　谢云玲　朱家鼎　谭炜骏　陈宇翔

概念原理 Concept

夏至之日，谓之朝节。熏风迎面，故古人常以彩扇相赠。民间也有"一九二九，扇不离手"之说。

影踱风舞——我们借助场地内扇形的长椅，插上木条制成椅背，并投射多彩的灯光，在地面上留下多彩的扇子投影，随风曼舞。周边的小灯笼则抽象出扇面的元素，多彩的灯光、重叠的图案，落在软质的随风漂浮的"外衣"上，点缀场地。通过控制光色，上下移动、旋转光源，通过光影互动展现夏至的热烈，扇子展开，扇子扇动，带来一阵清凉的夏至意境。

我们的最终装置从基地中自然生长出来，唯此一处可行，无法复制。

On the summer solstice day, wind blows, so in ancient times, people exchange fans as gifts.
Shadows' Wander With Wind - Using the premises benches, we create colorful fans on the ground, wandering with the wind. Little lanterns made of soft paper with colorful and different shadows of the pattern extracted from fan dott the area. We can modify unfolding or anning by rotating or moving the lights. Lights color changing between warm and cold means hot weather and cool fanning.
Our devices grow from the base, and only can be in this area. That is, it cannot be copied.

材料构成 Material

设计草图 Sketches

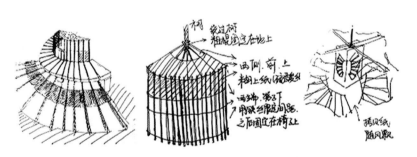

制作过程 Construction

成果表现 Realization

夏至·风曳雨荷
Lotus in the rain

作者 Designers

王挺　王宇泽　胡楠　吕欣欣　龚周隽堃　邵一恒

概念原理 Concept

我们小组的主题节气为夏至，地块为三好坞水上亭子及游廊。由于地块处于水塘中央，所以我们选取了夏日池中的荷叶作为原型，一方面在水中布置漂浮的荷叶，另一方面在游廊布置绑定 LED 光源的雨伞。水中的荷叶以泡沫为底板，以白卡纸为衬托，上面铺设光导纤维，用于表现荷叶叶脉的肌理。泡沫板上挖洞固定两种不同颜色的光源，长导线将开关引到岸上，方便控制。不同的荷叶使用不同颜色的光源，形成丰富变化。游廊上的雨伞均绑定三个原色的 LED 光源，分别独立开关，组合出多种光色效果。

Our group works on the Summer Solstice, and the region is on the pavilion of Sanhaowu in the campus of Tongji University. We choose lotus leaves, which represent summer, to fit the water-surrounded region. Lotus leaves float on the lake and umbrellas with lights are set along the gallery. The leaves are made of cystosepiment, cardboard, and light-guide fibre, to express the texture of lotus leaves. Two different colors of lights are set in the cystosepiment, and the switches are led to the shore by long guide lines, to simplify the control of light. Three primary colors are set on every umbrella, isolated to switch, which can make up various color effects.

材料构成 Material

 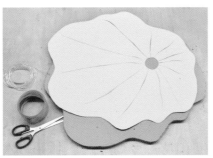

设计草图 Sketches

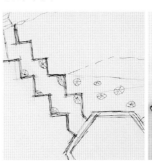  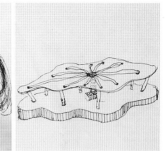

制作过程 Construction

     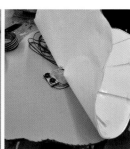

成果表现 Realization

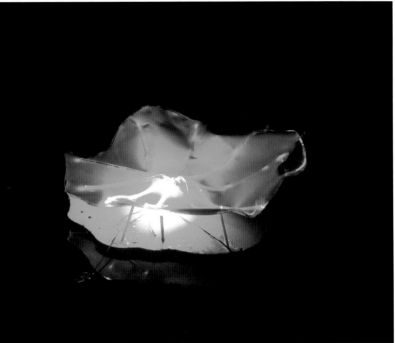

夏至·转一夏
Twist for the whole summer

作者 Designers

吴依秋　张琬舒　王子潇　胡淼　孙桢　王卓浩

概念原理 Concept

夏至时值麦收，自古以来都有在此时庆祝丰收、祭祀祖先之俗，以祈求消灾年丰。因此我们选择农村装谷物的竹匾为主要材料，通过旋转竹匾来象征庆祝丰收的仪式，以加强装置的参与性与互动性。我们装置的制作材料包括竹匾、麻绳、蛐蛐笼等。将两个竹匾倒扣为一组，在竹匾上挖洞，光线透过小洞在亭子的地与顶两个界面上同时形成梦幻光影，衬亮整个亭子。通过旋转，光影还会产生丰富的视觉变化。而同样用竹制成的蛐蛐笼则用来布置场地周边，引导视线。

The summer solstice is the time for wheat harvest. People celebrate the harvest during this time since ancient times, to worship the ancestors and pray for good harvest. So we choose bamboo plaques as main material, and simulate the harvest celebration ceremony by rotating bamboo plaques. Our materials include bamboo plaques, hemp ropes and cricket cages. Each two bamboo plaques are buckled as a group. Light passes through the digging holes on the bamboo plaques to form shadows on the ceiling and floor of the pavilion, lighting up the whole pavilion. By rotation, the light produces rich visual changes. The cricket cages are also made of bamboo to decorate the surrounding environment.

材料构成 Material

设计草图 Sketches

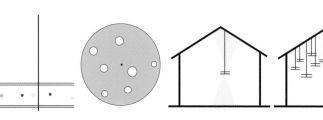

制作过程 Construction

成果表现 Realization

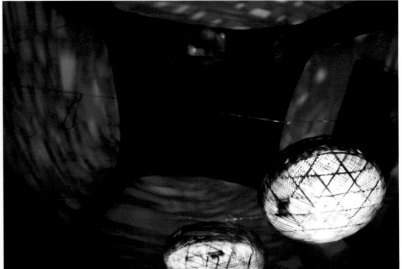

夏至·夏树
Summer tree

作者 Designers

蔡日革　安光成　金英进　李明成　魏 熊

概念原理 Concept

我们这次的光装置的主题节气是夏至。在每年公历 6 月 21 日或 22 日这天，白天最长，植物的新陈代谢最快，就是说植物开始旺盛起来的阶段。我们决定用光色冷暖来抽象地表达植物生长的过程，开始一段时间冷色的形状慢慢地成长起来，到了一定时间之后慢慢出现暖色，最后是冷色和暖色之间的随机变化。

Our theme is the summer solstice. Every year on June 21 or 22, the daytime becomes the longest in the year. During this time trees beign to florish. We decided to show a growth process of the abstact tree by switching between cold and warm colors. First the cold color's configration begins to grow gradually. Then it begin to appear warm color on the screen. Finally, the cold and warm color change mutually.

材料构成 Material

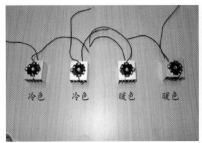 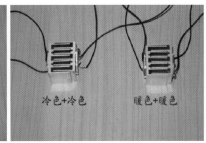

设计草图 Sketches

 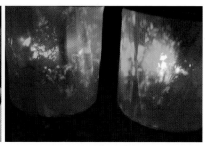

制作过程 Construction

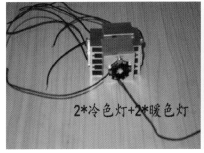   

成果表现 Realization

小暑・夏日梦舞
Summer daydream

作者 Designers

张晨阳　陶麒榆　王泰龙　程一　谈俊涛　朱方舟

概念原理 Concept

为充分表达小暑的特征，我们选择小暑时节开放正盛的荷花为意象，通过用不同颜色的花瓣作为生命的承载形式，将其投射在精心制作的承影面上，以表达昂扬而勃发的生命力。我们通过在牛皮纸上雕刻精心设计的孔洞，将其塑造成圆形灯罩。将精心挑选的焊接好的LED灯按照我们设计的方案安放在灯罩内，然后以硫酸纸为承影面呈现光影。装置的摆放最大限度地与基地结合，从而使映照在地面上的光影呈现出万物蓬勃生长的形态，充分反映小暑时节万木葱茏、千帆竞发的特色。

Lotus, which is at its prime during "Slight Heat", is chosen as the source of our inspiration. By cutting petal-like holes on krafts, we get a series of beautiful lanterns. Light from LED of various color filters penetrates through those lanterns, casting a glow as gorgeous as a summer daydream. Semitransparent drawing paper plays the role of the projection plane. Summer breezes blowing, drawing papers fluttering, petals dancing, and the summer daydream dreaming.

材料构成 Material

 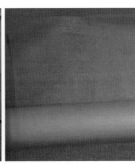 

设计草图 Sketches

制作过程 Construction

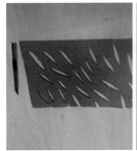     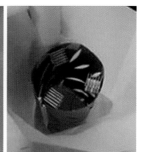

成果表现 Realization

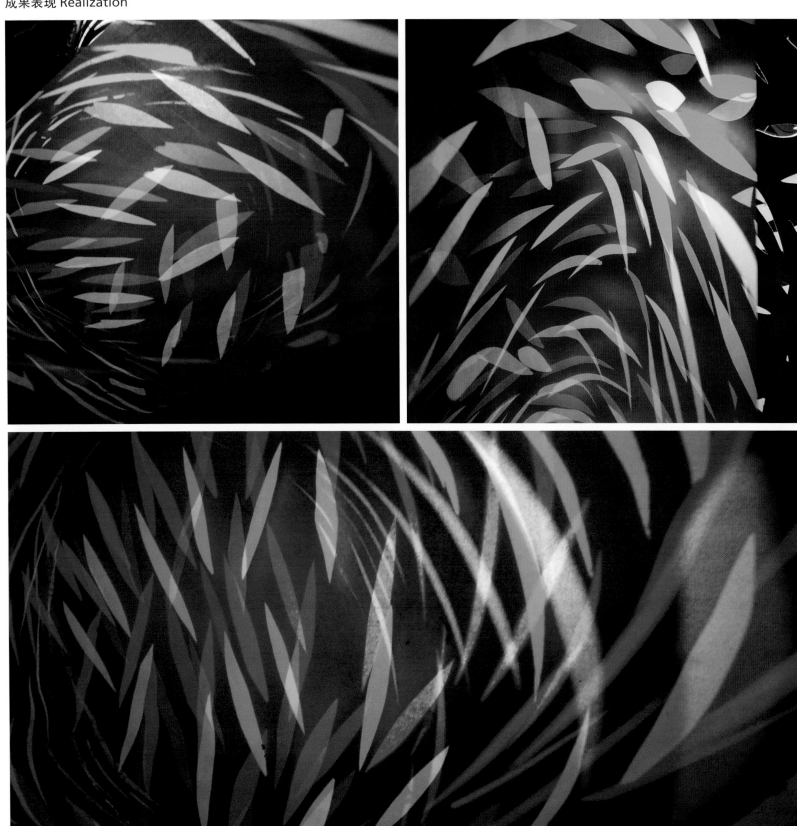

大暑・炽日裂岩
Hot sun breaking rocks

作者 Designers

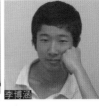

谭嘉琪　李博涵　郑思尧　王姚洁　程尘悦　郭佳鑫

概念原理 Concept

本小组的主题为大暑节气，大暑是一年中最热的时候。大暑让人印象最深刻的便是炽烈的太阳，在不下雨的时候，太阳的炙烤往往会致使地面干裂。我们小组即提取"太阳"与"地裂"两种意象，进行抽象与丰富，制作成为我们的光装置。"太阳"——用黑卡纸卷成筒，一侧用竹架固定，另一侧蒙上草图纸并面对观众，在筒内放置LED光源，筒边划出细长的洞，丰富光影效果，筒内加入一些自然元素；"地裂"——将可反光的纸裁出裂缝，贴在硫酸纸上，地面上放置LED光源，将装置的"地皮"铺在上面，则可形成地裂的效果，裂缝中有如熊熊烈火，突出"酷热"的特点。

The theme of our team is "Great Heat", which is the hottest time during the year. The most impressive feature at Great Heat is the fiery sun, while there is no rain, the ground is cracked under the burning sun. In that case, we collect the two images "burning sun" and "crack" to abstract them and rich them, making our light device. We do as follows: "burning sun"—— roll over the black card into a cylinder, with the bamboo cane fixing it on the one side, sketching paper covering on the other side facing the crowd. Put the LED inside the cylinder, line out long and thin slot on the face of the cylinder to rich the effect of light and add some natural elements in the cylinder; "crack"—— cut out cracks on reflective sheets and then stick them on the parchment papers. Lay LED under the parchment papers on the ground and the effect of suncracks will come out, as if buring fires in the cracks.

材料构成 Material

设计草图 Sketches

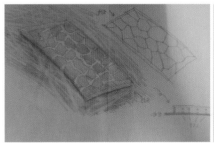

制作过程 Construction

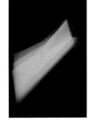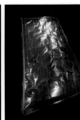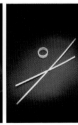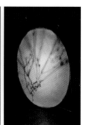

成果表现 Realization

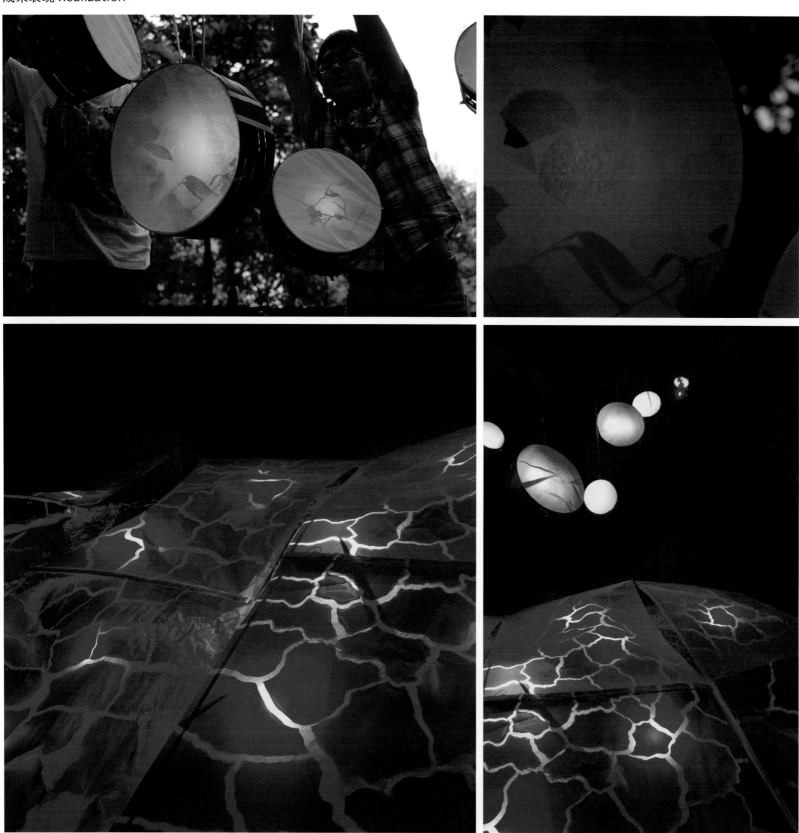

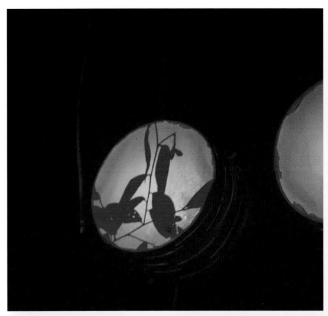

立秋 · 一叶知秋
The falling of one leaf heralds the autumn

作者 Designers

郑 馨　　张文易　　吴夏安　　陈容律　　姜睿涵　　金 鹊

概念原理 Concept

古语有云，梧桐一叶而天下知秋。立秋作为由夏到秋的标志性时节，虽然天气尚炎热，但是掉落的一片梧桐叶象征着秋天已悄然而至，树叶即将纷纷掉落，果实也即将挂上枝头。我们制作了很多刻有梧桐叶的灯笼，通过它们由少到多的点亮，来表现树叶缓缓掉落的过程。我们在灯笼的底部刻上散落的菱形、圆形等不同的形状。在灯点亮后，我们与观众一起掀起灯笼，使得底部的图案映射到地面、桌面上，呈现由静谧到绚烂的动态过程，营造落叶缤纷、果实累累的效果。

As the old saying goes, one falling phoenix tree leaf is indicative of the coming of autumn. The Beginning of Autumn is the symbol of transition from summer to antumn. Although the weather is still hot, a falling leaf shows that autumn has arrived secretly. Leaves are going to fall and fruits are going to drop off the trees. We made many lanterns with phoenix tree leaves carved on them, representing the increasing fallen leaves by lighting up more and more lanterns. We carved ruleless rhombuses or circles at the bottom of the lanterns and let the audience lift them to shade the pattern on the ground and the table. Thus, the audience would marvel at the emerging splendid scenery of colorful leaves and countless fruits.

材料构成 Material

设计草图 Sketches

制作过程 Construction

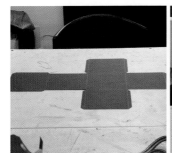 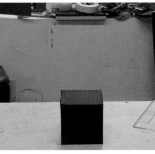   

成果表现 Realization

处暑·泛波竹海
Waving in bamboo ocean

作者 Designers

杨 竞　张 季　汤胜男　解李烜　王舟童　魏嘉彬

概念原理 Concept

该光学装置以透光的油纸伞作为载体，利用场地内具有丰富内部空间的竹林，将水纹灯按照流线变化逐渐由浅绿变为深绿再变为黄色最后变为紫色来配置其间，加以潺潺水声，营造出具有秋夏变化效果的禅意灯效，凸显"处暑"夏秋之交的节气特点。装置将光学与行为学结合，观者撑着油纸伞，伴着叮咚水声漫步在竹影阑珊的竹林间，感受着伞面上伴着潺潺水波的竹叶投影与光色变幻，体味处暑的节气意蕴。

This optical installation takes oiled paper umbrella as the carrier. By using the space of bamboo, we gradually change the color of our waving light from light green, to dark green, to yellow and then to purple, to make our lighting effect show the change of summer and autumn. At the same time, we utilize the sound of water to heighten the effect. We combine optics with behavioristics. When the visitor walks under our installation, the shadow of bamboo, the sound of water, and the change of lighting effect can make him/her feel the characteristic of the end of summer.

材料构成 Material

设计草图 Sketches

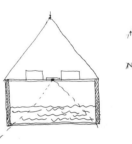  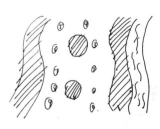

制作过程 Construction

成果表现 Realization

白露・白露满地
White Dew, the whole ground

作者 Designers

金广浩　李泰凡　金金日　朴英成　金哲成

概念原理 Concept

我们这次的光装置的主题节气是白露，在每年公历9月7日或8日这天。露是由于温度降低，水汽在地面或近地物体上凝结而成的水珠。所以，白露实际上是表征天气已经转凉。这时，人们就会明显地感觉到炎热的夏天已过，而凉爽的秋天已经到来。因为白天的温度虽然仍达三十几摄氏度，可是夜晚之后，就下降到二十几摄氏度，两者之间的温度差达十多摄氏度，所以我们决定以叶子上面的露珠为意象，用不同的颜色给人很清凉的感觉。为了达到这样的目的，我们采用的材料是黑卡纸与防震膜以及药片（环保的目的），考虑到白露季节的温度变化，我们在桥梁上做出了一些颜色过渡（从暖色到冷色）。此外，我们还为装置设计了自动控制系统。

Our theme is "White Dew", the day on September 9th or 10th. Because of the low temperature, the moisture on the ground or on a near-Earth objects condense into droplets. This feast means the weather has already been cool from that time on. Although the temperature in daytime is up to 30 celsius degree, in the nighttime the temperature will decreases to 20 celsius degree. According to this characteristic, we decided to describe the dew on the leaves to let people feel cool by creating different colors. To achive this goal, we use black cardboard, earthquack film and tablets. What's more, we made some color-changing on the bridge to represent the temperature-dropping. Besides, the automation system was used to control the unit successfully.

材料构成 Material

设计草图 Sketches

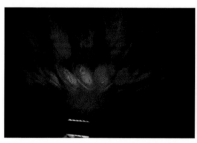 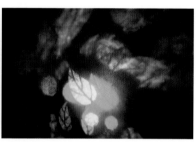

制作过程 Construction

 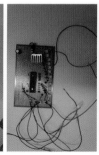  

成果表现 Realization

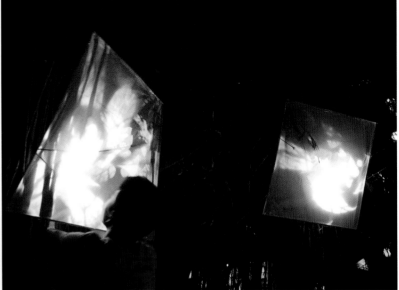

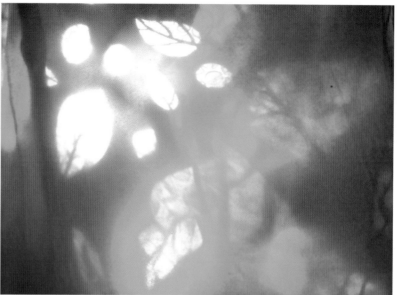

秋分・雁来燕去
Waltz of winged miration

作者 Designers

辛诗奕　　贺艺雯　　王翱　　唐楷文　　张若松

概念原理 Concept

本小组的主题是秋分。
形式来源：候鸟迁徙。
主题诗：白露秋分夜，雁门开、雁儿脚下带霜来。燕将明日去，对夕曛、秋分种麦正当时。
基地结合：水上水下对称，呼应昼夜平分。基地狭长，水中倒影，影姿绰绰。
整体色调：金黄色、琥珀色、浅黄色、亮白色。

The theme of our group work is "Autumnal Equinox".
Source of manifestation: Migration of birds. Swallows are flying to the south district while wild gooses are coming back when autumnal equinox is approaching.
Poetry: It results from a Chinese poetry which describes the beauty of the autumnal equinox. It shows the magration of the birds to imply the change of seasons.
The idea in accordance with the environment: The rivulet is narrow and long. The fabulous and fantastic light of swallows and wild gooses and its reflection in the rivulet stands for equinoctial.
The tone of the whole light work: golden, amber color, pale/light yellow, together with the bright white.

材料构成 Material

设计草图 Sketches

基地布置　　　　　　　　　　　白　琥珀　黄

制作过程 Construction

   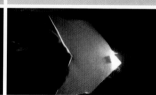

成果表现 Realization

寒露·白露为霜
Dew cream dream

作者 Designers

王劲凯　赵月僮　杨舒丹　安麟奎　丁种铉

概念原理 Concept

我们做的主题是二十四节气中的寒露。
寒露在每年的10月8号或9号，与重阳节相临近，此时天气转寒，雾气即将凝结成霜，菊花绽放。因此，我们想要以"白露为霜，菊有黄华"来表现主题，希望人们在廊桥上行走，仿佛踏于白霜之上，菊花映目。在这炎热的夏季能够感受到深秋的清寒，同时又仿佛能嗅到菊花的芳香。
光装置分为两个部分，采用了冷暖光对比的手法。
冷色光部分，割开纸板，进行折叠，形成凹凸有致的方格，在其间放入冷色 LED 光源，以营造片片散落白露成霜的氛围。
暖色光部分，运用瓶子、竹竿、黑色雪弗板等，灯光透过瓶子打在黑色的板上，可以清晰地看到菊花的形状。

The theme of our light installations is the cold dew.
Cold dew comes on October 8 or 9 every year, with the Chung Yeung Festival approaching. When the weather turns cold, the fog condenses into frost, and chrysanthemum bloom. Therefore, with the theme: Dew condenses into frost, and chrysanthemum bloom, we want to reach the effect that people walk in the corridor as if riding on frost, with chrysanthemum mapping projects. In this hot summer visitors can feel the raw cold of late autumn, while smell the fragrance of chrysanthemum.
The light installations are divided into two parts, using cool and warm light, to obtain contrast.
Cool light parts, cut parts, folding, forming convex grid, to create a piece of Bai Lucheng scattered frost atmosphere.
Warm light section, the use of bottles, bamboo, black Chevron board, etc., through the bottle hit the lights on a black board, you can clearly see the shape of chrysanthemums.

材料构成 Material

设计草图 Sketches

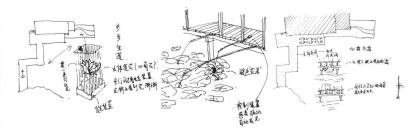

制作过程 Construction

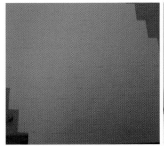    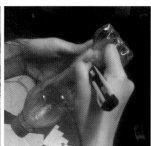 

成果表现 Realization

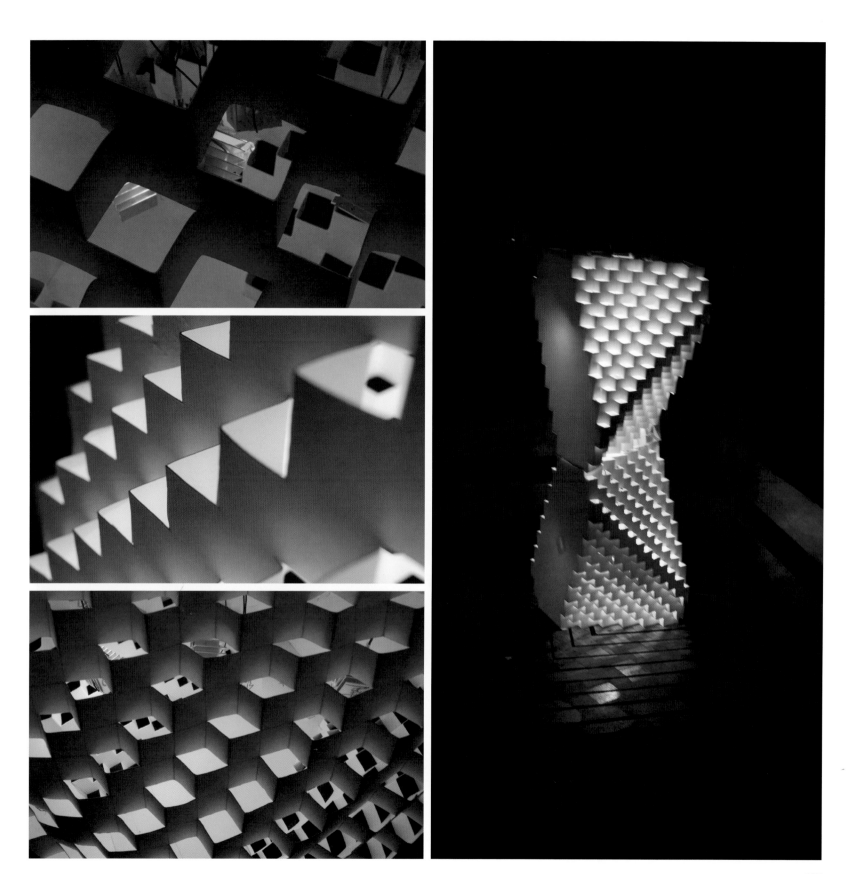

霜降·落幕飞霜
Blue blue glass moon

作者 Designers

  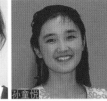   

王 敏　姚懿芸　孙童悦　林静之　杨鹏程　石彦杰

概念原理 Concept

霜降——秋末冬至之际，万物凋零。光装置灵感源于电影《英雄》片段。我们刻制了夏、秋、冬三种花纹的木扇，将其挂于河面之上，用冷光（蓝、白两色）打出扇影在波光粼粼的河面上，犹如落霜。另外，用暖色的蜡烛（通过橙色塑料吸管和一次性纸杯组合制作而成）来烘托飞扇的肃杀，制造祭奠的意味。营造出风吹残烛、落幕飞霜的凄冷氛围。

The idea originated from the concept of "First Frost". In traditional Chinese culture, "First Frost" marks the end of autumn and the beginning of winter. It is a season where nature begins its slumber, a time of whitering and death. This "ending" is expressed through light and shadow cast through lattices imprinted on the fan-shaped filters. The lattice illustrates the transition from autumn to winter. A ceremonial atmosphere is brought out by the use of "candles", which transforms the site into an altar telling the story of "First Frost".

材料构成 Material

设计草图 Sketches

制作过程 Construction

成果表现 Realization

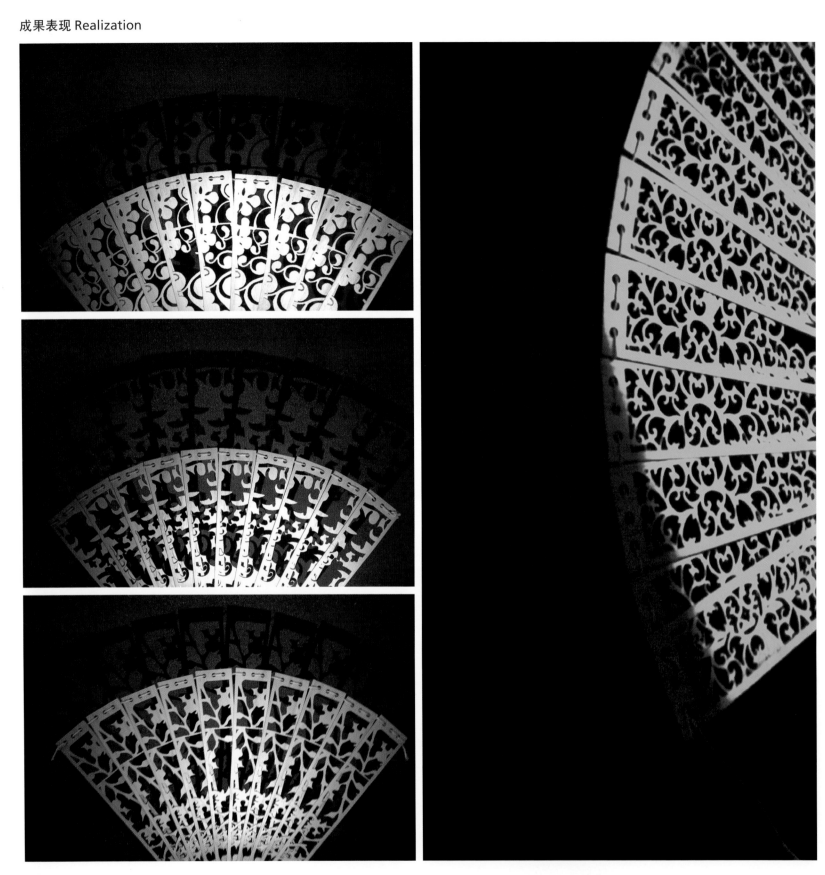

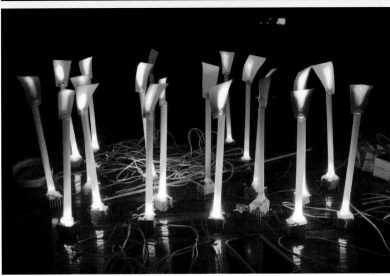

立冬
Winter Begins

作者 Designers

   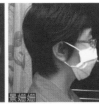  

杜叶铖　陈元　王钦　景姗姗　叶之凡　程叙

概念原理 Concept

立冬是冬天的第一个节气。从"立冬，万物收藏也"汲取理念，做若干不同大小、不同边长数、表面糊一层拷贝纸的棱柱，再做一些刻有谷物图案的边长为5厘米的小盒子，将光源置入其中。光源透过盒子将谷物的影子投射到棱柱上，用光将营造出"万物收藏"感觉。水则用纯色的河灯代替。因基地有两级阶梯延伸到水中，所以从阶梯到水中的光色从暖到冷，营造出一种由秋天到冬天过渡的氛围。

"Winter Begins" is the first jieqi in winter. We draw our concept from an ancient Chinese sentence: winter begins and keeps everything in storage. So we decided to make some prisms with different sizes and the length of side. Tissue paper were pasted on the surface of the prisms. Then we put LED lights into a small box whose size is 5×5×5cm³ with patterns of grain. Light is found on the tissue paper through the small box inside. That's what we said "Everything in Storage". On the water, we use river light with pure color. Because our base consists of two stairs and water area, we decide to use warm color on the first stair and cold color on the water to create an atmosphere where we see autumn off to welcome winter.

材料构成 Material

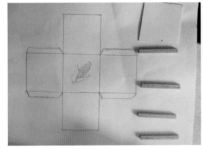 

设计草图 Sketches

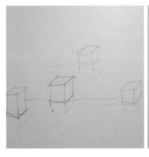  

制作过程 Construction

    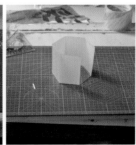 

成果表现 Realization

小雪
Light Snow

作者 Designers

王浩然　闫爽　何美婷　于越　李霖　戴方睿

概念原理 Concept

光装置——"小雪"渲染了初雪从伞间洒下，随风飘荡，慢慢坠落到地面融化的浪漫气氛。人在装置之中穿梭行走、欣赏拍照，如置身初雪之中。该装置灵感源于对小雪的理解——清新、轻盈、透亮，在空中随风飘散，于地面融化消失。故先将反光纸剪成小块、打孔，穿在细鱼线上，系在下雪时使用的伞的伞骨上，模拟忽暗忽现的雪花。在伞内侧顶端挂上固定有 LED 灯的塑料瓶。瓶子的凹凸纹理在光的作用下投射出点状光斑，如雪花点点。伞悬挂在空中呈漂浮状。地面放置以竹条为框、白色垃圾袋为面的盒子，内部放置连有电容的 LED 灯，呈现忽明忽暗的效果，模拟雪在地面融化的感觉。

The lighting device called "Light Snow" creates a romantic atmosphere and imitates the first flurries of snow drifting through the umbrellas, blowing in the wind and light and noiseless as it floats down. People can walk through the device, taking photos while appreciating and feeling that they are standing in the snow. The device is inspired by the understanding of the light snow which is pure, fresh, light and bright. It gets blown away by the wind and melting on the ground. The reflective pieces are floating in the air like the flurries of snow. The LED lights inside the on the ground boxes create a flickering atmosphere and imitate melting snow.

材料构成 Material

制作过程 Construction

成果表现 Realization

大雪
Great Snow

作者 Designers

陈珂怡　　郭欣　　景巍然　　吴风　　张弛　　朱元元

概念原理 Concept

大雪这个节气总能让人想到"瑞雪兆丰年"这样的吉祥话。漫天飞舞的雪花不仅是天空的使者，更是世人美好愿望的载体。

我们的灵感来自北欧雪国，想营造出一种温馨、静谧而甜蜜的氛围。我们利用锡箔多角度反射出的光斑来表现雪花；大面积铺设在基地上的拷贝纸和棉花帮助我们营造出宏大的场面；扇形的基地和上方有规律铺设的三角形小房子夸大了近大远小的透视效果。我们并没有采用惯常的冷色调来表现寒冷和雪景，反而是以温暖的黄橙色调为主，希望可以着重表现大雪带给人们的温馨氛围。

屋外的雪有多大，家就显得有多温暖。

The solar term of Great Snow always reminds people of the Chinese saying "A fall of seasonable snow gives promise of a fruitful year". The flying snow is just like the messenger from the heaven who carries the sweet dreams of people. Our idea comes from the countries in Northern Europe and we hope to create a peaceful, sweet atmosphere. We use tinfoil to represent snow with its reflection of light in multi-direction. And we also build a spectacle snow scene by spreading the base with tissue paper and cotton. On the paper, we exaggerate the feeling of perspective, that is, we put small houses in the back and big ones in the front. In a whole, the reason we use warm colors for the lights is also to intensify the feeling of warmness in a snowy day.

材料构成 Material

设计草图 Sketches

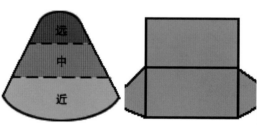 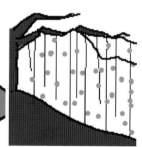

制作过程 Construction

   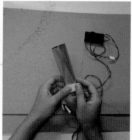  

成果表现 Realization

冬至
The Winter Solstice

作者 Designers

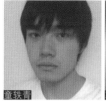    

童轶青　郭子豪　张雨缇　何绍禧　李月光

概念原理 Concept

我们组做的节气是冬至。在小组讨论初期，我们总结了冬至中国的习俗。大部分地区的中国人都要在一起吃饺子度过冬至，迎接冬天的到来。"团圆"这一意象也被我们融入了设计之中。同时，冬至这一天也是太阳直射南回归线的一天，北半球在这一天白昼最短，但在其中却蕴含了白昼由短变长的契机。事物的循环往复、兴衰生灭也成了我们设计的一个依据。

有趣的是，我们组有一个缅甸的留学生同学。他提出了富有特色的理解。由于缅甸位于热带地区，雨林丰富，蜗牛这一生物对于缅甸人来说具有非同寻常的意义。生活中处处可见的蜗牛不但是一种美味的食材，也是生活的一个符号：在每年的末尾，缅甸人也会聚集在一起庆祝一年的结束。然而冬天是没有蜗牛的，蜗牛形状的食物也成了缅甸人的重要食物。

The solar term we work on is "Winter Solstice", which means "winter coming" in Chinese. During the first discussion, we made some conclusion on the Chinese customs of Winter Solstice. People from most regions of China eat dumplings on this day. As a result, we put the concept of reunion ("Tuanyuan" in Chinese, which also means a round circle) into our lighting design. On this day, day time in the northern hemisphere is the shortest, after that, the day starts to get longer. The cycles of things, the birth and death rise and fall of a basis also became our design concept.

It is interesting to note that there's a foreign member from Myanmar in our group. As Myanmar is from the tropical region, where it's rich in rainforests. Snails are treasured creatures, which are not only tasty food but also a symbol of friends and family gathering to celebrate the end of the year. However snails are few in winter, so food shaped like snails are popular in Myanmar in winter.

材料构成 Material

设计草图 Sketches

制作过程 Construction

成果表现 Realization

小寒・浮月寒冰
Song of moon & ice

作者 Designers

王宇昊　　崔芳毓　　杜伊卓　　黄翊宁　　阳腾飞　　朱君燕

概念原理 Concept

我们的光学装置作品主题为二十四节气中的小寒，一年中最为寒冷的时节之一。小寒是滴水成冰的日子，我们便以这个特点为创作基础，在装置中选用蓝白两色的LED灯表现水和冰。上方三束用玻璃丝包裹的蓝色LED灯代表水流，展示时以动态顺序点亮，模拟水滴落下的过程。下方小装置均是用硫酸纸和珍珠棉包裹LED灯做成，三盏蓝灯代表落入水面的三滴水，环绕的白灯则是冰块。装置整体下层铺有反光的塑料纸，营造出装置下部水面波光粼粼的环境。装置整体呈冷色调，渲染出小寒时节冰冷萧然的气氛。

The theme of our optical device is "Slight Cold", one of the twenty-four solar terms. "Slight Cold" is one of the most freezing time in a year, when water drops can easily become ice. Based on this character, we choose white and blue LEDs to imitate water and ice in our optical device. Above the device, we use three beams of glass silk wrapped blue LED lights to represent flow. When displaying our device, the LEDs will be lighted in a certain order to imitate the falling of water drops. The part of device on the ground is made of paper and cotton wrapping LEDs. The three blue lights represent the three drops faling into the water, surrounded by the ice which are made of white lights. The ground is covered with plastic paper, creating a reflective water circumstance. The whole device shows a cool tone, which can foil the atmosphere of slight cold.

材料构成 Material　　　　设计草图 Sketches

 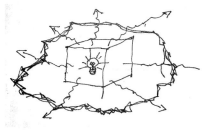 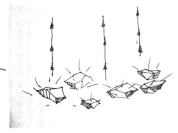

制作过程 Construction

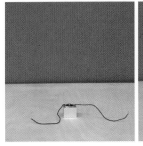     

成果表现 Realization

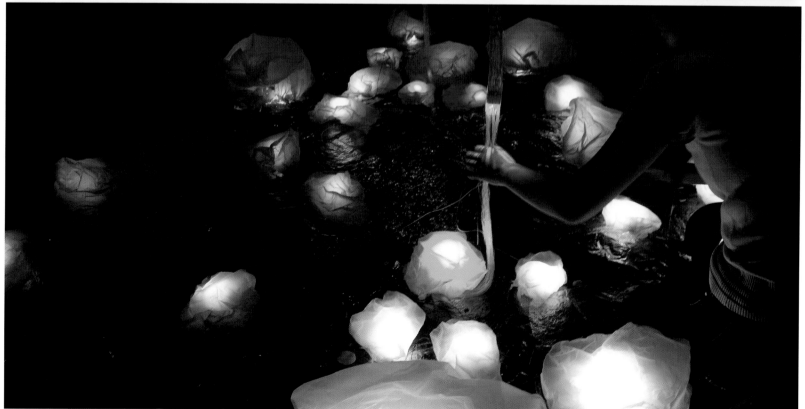

大寒
Great Cold

作者 Designers

赵曜　姜鸿博　陈嘉禾　苏南西　徐濛　洪逸伦

概念原理 Concept

大寒是二十四节气中的最后一个节气，是中国大部分地区一年中最冷的时刻。通过此次灯光展示，我们想要在炎热的夏季中创造冰天雪地的氛围。我们提取大寒的特征——冰、雪，作为我们灯光装置的主体元素。利用塑料桌布包裹基地中的石块等现有环境，并把LED灯和装满清水的玻璃瓶组成的装置置于塑料桌布之下，使灯光散射混合在桌布表面，以达到大体量冰块的效果。将反光贴纸裁成小条，成串挂在树梢，反射锥形装置发出的点点灯光，模拟雪花的效果。我们设想的整体效果是，雪中大地开始慢慢结冰，寒风更甚，达到最冷的时刻，而就在满目风雪时，忽然之间，从中透出点点暖意，预示着春天的来临。

Great cold is the last term of the 24 solar terms, it is also the coldest moment of the year. By displaying the lights, we want to get the feeling of ice and snow. These two main features of great cold form the elements of our installation. Plastic tablecloths are used to wrap the stone and other surroundings in the site, under which we place LED and glass bottles filled with water. In this way, the lights are mixed and scattered on the tablecloths, which looks like giant ice. We cut the reflecting paper into strips, stick it onto the fish wire and hang it on the branches so that it can reflect the light from the cone lantern with the trees to imitate the snow. Finally, we use glass installation to transit the ice to snow. The scene we imagined is that the snowfield start to freeze, the wind blows stronger and reaches the coldest moment. At this moment, warm light comes out, indicating the spring.

材料构成 Material

设计草图 Sketches

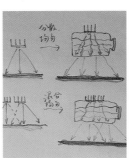  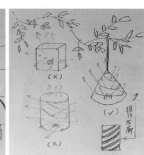

制作过程 Construction

     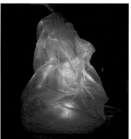

成果表现 Realization

138

立春・解冰曲水
Ice melts into serpentine water

作者 Designers

邹天格　　许双盈　　tanny　　吴垣锡　　吴汉祥　　万远超

概念原理 Concept

"一候东风解冻，二候蜇虫始振，三候鱼陟负冰"——我国古代将立春的十五天分为三候。从东风解冻，到幼虫振翅，再到鱼潜碎冰，是一个生命力逐渐由束缚到解放的过程，是一个慢慢变化的动态伸展。立春象征着在束缚中的具有生命力的动态变化。为了在光艺术设计中体现立春节气在束缚中的动态生长与变化，从"东风解冻，鱼陟负冰"中提取出冰的角色，代表了过去与束缚，再通过冰裂到融化和颜色渐变的过程，利用投影的方式体现立春的动态生长与变化。

The beginning of spring in ancient China, is divided into three periods. The fisrt period is when east wind comes, ice then melts. The second is when insects begin to fly. The third is when fish start to swim under the ice. This is a process vitality gradually from bondage to liberation and slowly changing dynamic stretch from the first period to the third. Beginning of spring symbolizes the dynamic changes in bondage with vitality. In order to reflect the beginning of spring solar terms in the light of Art and Design in bondage dynamic growth and change, from "East wind coming , ice melting and fish swimming under the ice" role in the extraction of ice, on behalf of the shackles of the past and then to through the ice cracking and melting and color gradual changing process, we use approach of projection to reflect the dynamic growth and change of the beginning of spring.

材料构成 Material

设计草图 Sketches

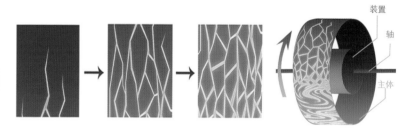

制作过程 Construction

成果表现 Realization

雨水・雨的抒情
The rain lyric

作者 Designers

 伍曼琳
 沈依冰
 王 林
 赵艺佳
 潘 屾
 王康富

概念原理 Concept

用灯光模拟下雨的景象,并配合真实的水滴来迎合我们的主题"雨水"。我们将废旧的易拉罐容器作为聚光以及盛水材料,将塑料管作为导光以及导水管连接于易拉罐下方。这样不仅导管本身发光,导管下方的地面也会有类似水波涟漪的投影产生。此外,我们还利用矿泉水瓶本身的纹路制造大的涟漪投影到地上。我们组的基地是一座桥,为了实现场景,我们在桥上设立架子,再将并联的易拉罐灯放置其上,布成阵。最后配合音乐控制开关,让不同组的灯跟着音乐节奏开启关闭,使之更为生动。

To create a rainy scene by using the lights and real water drops, we use wasted cans as condenser materials and water containers, and put plastic pipes underneath to let light and water come out, to create a beautiful shadow of ripple on the ground. We use plastic bottles to create bigger ripples as well. In order to make our installation, we made a frame and put it on our bridge as base. We hung the lights in line and switched them with the rhythm of background music.

材料构成 Material

设计草图 Sketches

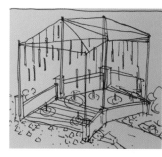 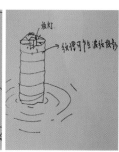 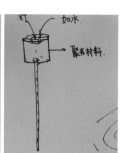

制作过程 Construction

   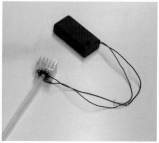 

成果表现 Realization

惊蛰
The waking of insects

作者 Designers

申艺振　　纳斯佳　　瑞 驰　　汉诺克　　路 烁　　阿 谷　　微 亚

概念原理 Concept

我们组的主题是"惊蛰"，这是一个冬天离开、春天到来的节气。 我们在辞典上查了"惊蛰"的意思。原来很久以前，人们觉得雷电划过，雷声太大，虫子受到惊吓，都醒来了，万物复苏，于是春天开始。
我们想表现这种感觉，所以从冬天开始，下雪了，刮风了，什么都没有，就是冬天。突然电闪雷鸣，吃惊的虫子都出来了，蝴蝶飞了，萤火虫闪来闪去，花开了，青蛙蹦了，春天来了。

The theme of our group is "the waking of insects", the period when winter is leaving and spring is coming. We were searching for what it meant and we found that in actient times, people thought when there is a thunderstorm, insects are surprised and awake and so they come out to the earth then spring starts. We tried to express that so we started from winter first. It is snowing and windy, nothing is moving, just cold winter. Then thunderstorm with lightening come and everything is awake; butterflies are flying, firebugs are emitting lights, flowers are blooming, frogs are jumping around and finally spring comes.

材料构成 Material

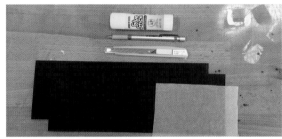 

制作过程 Construction

成果表现 Realization

150

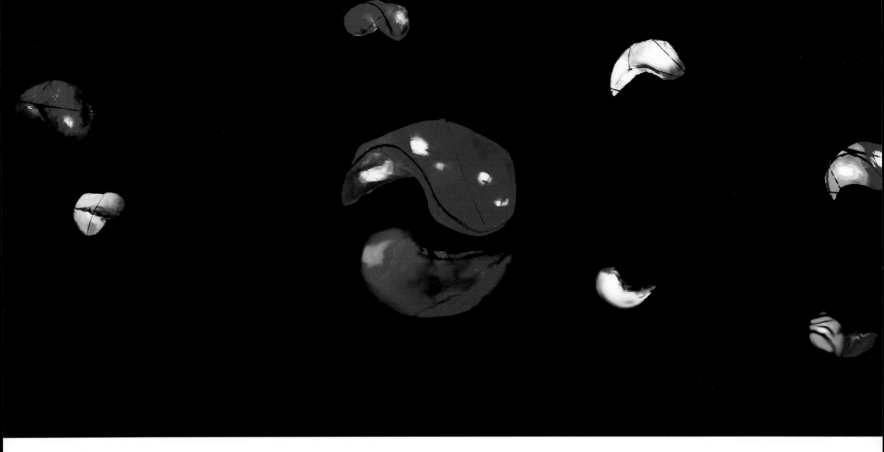

春分·春春鱼动
The spring Equinox
yin-yang fish dance

作者 Designers

樊怡君　　李萧文　　沈若玙　　徐　亮　　肖子颖

概念原理 Concept

"春分者，阴阳相半也，故昼夜均而寒暑平。"平衡在中国古代二元论的世界观中的具象表现形式是阴阳鱼图；同时根据基地邻水的特点，使阴阳鱼成为设计的概念，装置利用单体的阴阳鱼形状，和河面镜面反射的原理及透视原理，经过视角的计算之后进行造型，形成立体阴阳鱼形的抽象形态。此形态的特点是：其立面和其在河水中的倒影并非关于河面轴对称，而是一虚一实的阴阳鱼形态，组合起来可以形成一个完整的太极图，以此来表达春分阴阳相调和的寓意。

"Chinese vernal equinox means 'Yin' is equal to 'Yang' as a half. So both day and night, cold and heat are balanced." The pattern of Yin-yang fish is the representational form of balance in the world, from the view of ancient Chinese. Our site is at the bank, which lead us to the form of fish. We use Yin-yang fish as a source of concept. We calculate the angle based on the principle of reflection and the principle of perspective. Finally we figure out a three-dimensional abstract form of Yin-yang fish. It's characterized by the relationship between its facade and its reflection in the river which together combined to form a complete Taiji diagram.

材料构成 Material

设计草图 Sketches

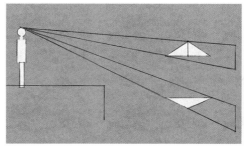 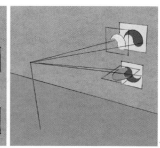

制作过程 Construction

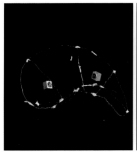 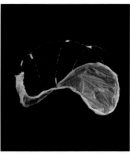 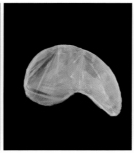  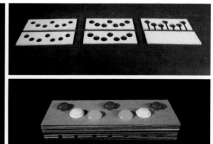

成果表现 Realization

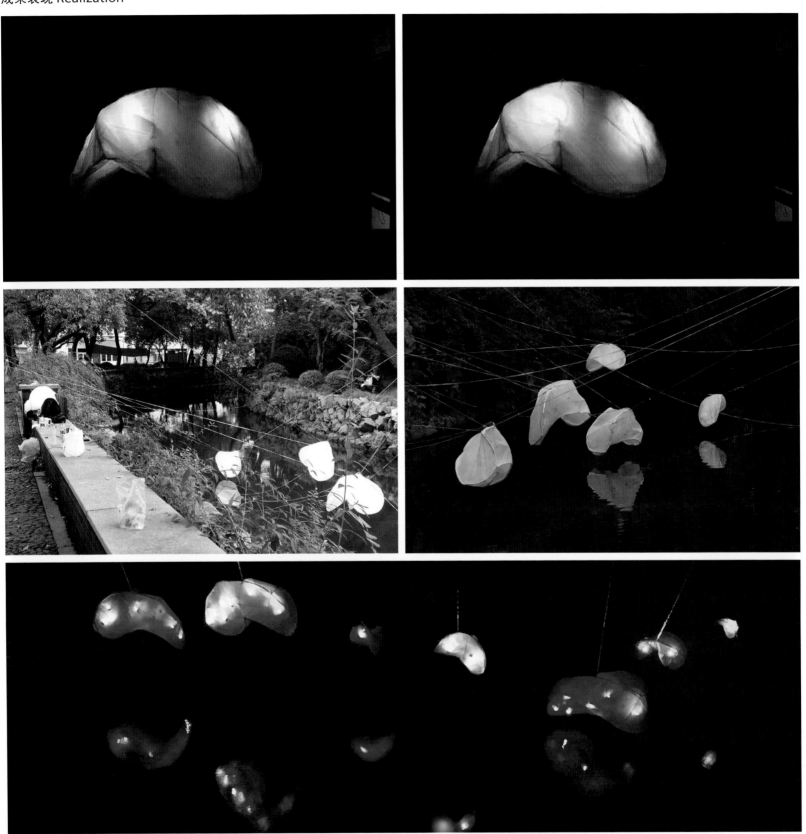

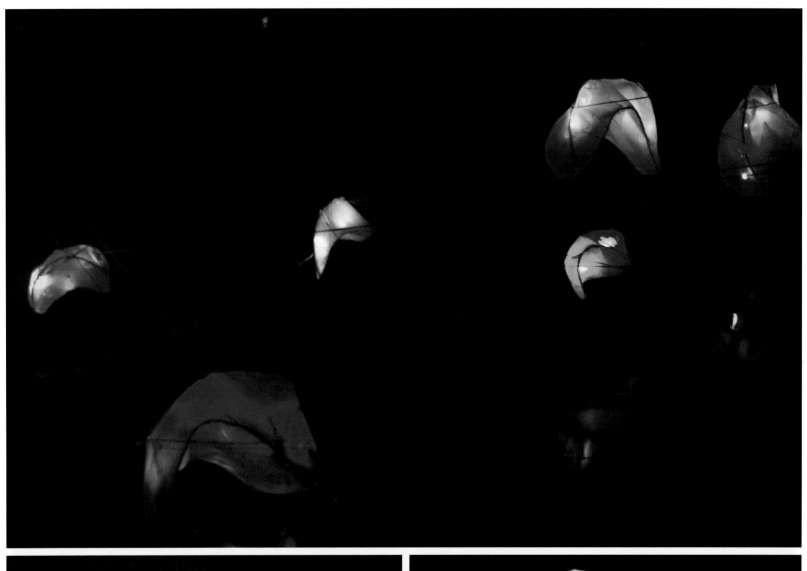
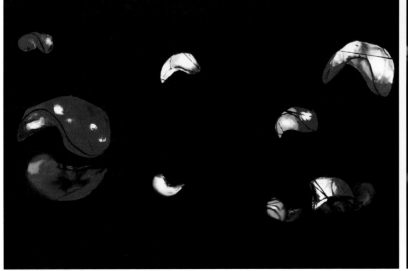
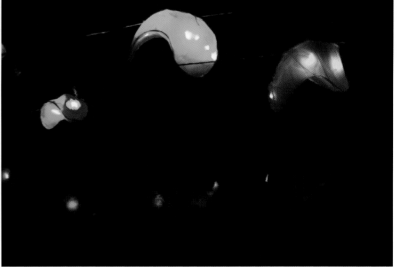

清明·蔚云细雨
Cloud and rain in Qingming

作者 Designers

朱佳桦　王韵然　李曼竹　胡秋叶　郝伟勋　谢天伟

概念原理 Concept

一片低垂的蔚云
几缕微凉的细雨
缱绻情思　流连怀恋
消散于这一片清明云雨间
以浴帽包裹灯光，形成一片色彩柔和朦胧的云，不同的开关控制不同的色彩，开关的开合形成丰富的闪烁和变化。被吸引而来的路人也可以参与控制灯光的变化，在婉转的音乐之中与身边或熟悉或陌生的人共同创造光的美景，体验会心一笑的默契。

Lights were wrapped in a shower cap, forming a hazy pastel clouds. Different switches control different colors, with different people control the on and off switches to form an interesting flicker and change. Passengers attracted by the scenery can also be involved in the control of changes in the melody of the music, staying with familiar or unfamiliar people together to create a beautiful light scenend experience a understanding smile.

材料构成 Material

制作过程 Construction

   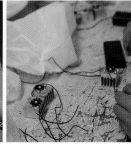  

成果表现 Realization

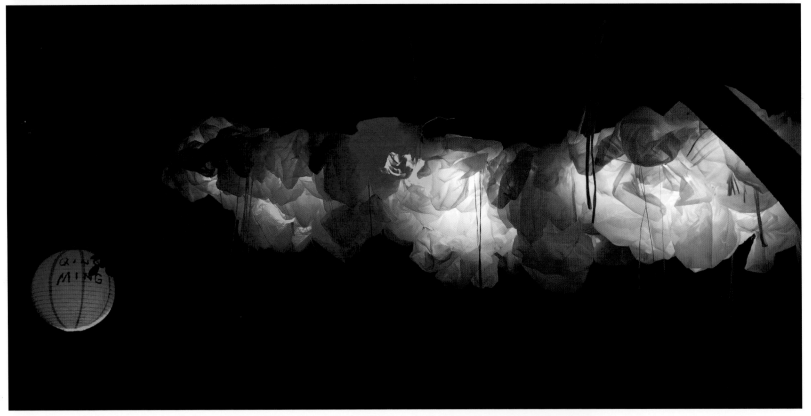

谷雨·生长时节
Flourishing season

作者 Designers

梁以伊　马慧慧　白珊山　戴方国　程晏宁　张鹏翔

概念原理 Concept

谷雨是春天的最后一个节气，代表着万物生长。我们小组以基地中的竹子作为生长元素，利用两个纸杯中间透明胶带模拟竹节，以开关控制每个"竹节"亮起来的时间。最后利用宣纸作为承影面，利用透明胶带对光的透射及散射性能，在宣纸上投射出竹叶的形象，呈现一个竹子完整的生长过程，做到了合情合景。

Grain rain, the last solar term of spring, represents the growth of living things. We chose bamboo as the typical element of growing, ultilizing transparent tape between two paper cups to emulate bamboo joint, manipulating every bamboo joint with seperate switch. At last, we use rice paper as our easel plane, considering the reflection and scatter capacity of the transparent tape, reflecting the shape of the bamboo levels. The whole performance represents a complete process of bamboo growth.

材料构成 Material

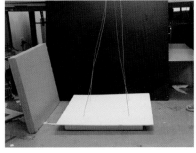 

设计草图 Sketches

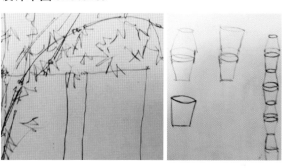

制作过程 Construction

成果表现 Realization

教学花絮 Teaching highlights

图1 图2 图3 图4

图 1、2、3、4
展示前各组学生在三好坞搭建场景。

图 5
研究生姚懿芸向学生们讲授并示范 LED 焊接。

图 6
学生们亲自动手尝试 LED 焊接。

展示颁奖 Show awards

图 1
任课教师宣布获奖作品。

图 2
悉尼大学教授 Warren Julian 为获奖学生颁奖。

图 3、4、5、6
获奖小组上台领奖并发表感言。

历年教学回顾 Review of years of teaching

| 2002 | 2003–2005 | 2006–2007 | 2008 |

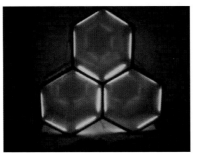

光与水立方——北京奥运会国家游泳中心（水立方）室内及立面光环境设计

课程要求学生对生活中的光环境进行调研，利用实验室的设备，采用不同的材料和照明方式，截取有意思的片段进行表现。这是本课程体系首次尝试脱离常规的教学模式，让学生亲自体验光照场景，理解物理量的视觉意义，以培养他们运用所学知识和设计技法解决实际问题的能力。

西安汉阳陵遗址保护展示厅及周边环境照明设计

在建筑光环境实验性教学体系探索初期，光艺术装置教学是照明专门化毕业设计中的重要环节。学生们根据自己的方案制作整个建筑或局部片段的缩尺灯光工作模型，在建立光环境概念的同时，掌握第一手光度数据资料，更好地将光度物理量与实际效果相联系。

利用传统光源如白炽灯、卤素灯、荧光灯等，易于市场选购的光源和实验室的设备，将光作为生动的设计语汇进行创作。其形态、光影、明暗构成动态或静态的光照图示。学生最终将成果制作成幻灯片或多媒体动画，展示他们对光与材质、空间的认识和理解。

由于照明技术的发展，光艺术装置实验环节引入低碳环保的LED光源作为创作素材。教学场地的LED显示屏成为同学们的创作基础，每组学生分配25cm×25cm大小的显示屏，他们在充分了解LED发光特点、材料特性的基础上进行光艺术媒体界面的创作。

| 2009 | 2010 | 2011-2012 | 2013-2015 |

2010世博会前期的学生作业,在世博文化中心两个入口大厅背景墙得到深化应用,最终作品完全由师生现场制作完成,获得了应用单位的高度认可,取得了良好的社会反响。并在若干年后被应用于实验室人居空间健康照明的研究课题中。

课程更加注重实践性,学生亲自动手焊接LED颗粒并连接电源。装置作业的规模、成熟度及艺术效果得到了加强。并在最后加入评审展示的环节,评审老师根据展示现场效果给出课程成绩。

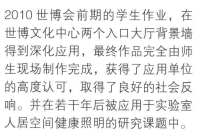

"声光SHOW"大型装置中,学生们在了解语音输入、声音信号识别、人机交互等原理的基础上,自己编排声光效果,再利用矿泉水桶作为发光像素点,上演了一场别开生面的"声光秀"。

从2013年开始装置作业有了主题,学生们在限定主题、场地、时间、装置制作材料、LED颗粒颜色和数量的条件下,以小组为单位,制作光艺术装置,结合"徒手控制"的动态编排,探索光影如何与空间、环境、音乐互动,探讨光与色彩如何改善生活环境、诠释传统文化、演绎戏剧情节。发现光的多维度应用,体验光的无限魅力。

图书在版编目（CIP）数据

2014·光影时节 / 郝洛西等著 . -- 上海：同济大学出版社，2016.6
（同济大学建筑学专业"建筑物理（光环境）"教学成果专辑）
ISBN 978-7-5608-6200-2

Ⅰ.①2… Ⅱ.①郝… Ⅲ.①建筑物理学—教学研究—高等学校—文集
②建筑光学—教学研究—高等学校—文集 Ⅳ.① TU11-53

中国版本图书馆 CIP 数据核字 (2016) 第 025195 号

同济大学建筑学专业"建筑物理（光环境）"教学成果专辑
2014·光影时节

郝洛西 等 著

责任编辑	张　睿
责任校对	张德胜
装帧设计	李　丽
APP 制作	今尚数字
出版发行	同济大学出版社（www.tongjipress.com.cn）
地　址	上海四平路 1239 号（200092）
电　话	021-65985622
经　销	全国各地新华书店
印　刷	上海丽佳制版印刷有限公司
开　本	787mm×1092mm　1/12
印　张	37.33
字　数	940000
版　次	2016 年 6 月第 1 版
印　次	2016 年 6 月第 1 次印刷
书　号	ISBN 978-7-5608-6200-2
定　价	210.00 元（全三册）

○ 向 2015 国际光年致敬

同济大学建筑学专业"建筑物理（光环境）"
教学成果专辑

2013·南极之光
THE LIGHT OF THE ANTARCTIC

郝洛西 等 著

同济大学 出版社
TONGJI UNIVERSITY PRESS

目录 Contents

Part 1		Prologue 写在前面	6
Part 2		Task introduction 作业简介	22
	01	Antarctic window, colorful cube 南极之窗，炫彩立方	24
	02	Guang guang guang light 咣咣咣光	28
	03	Golden days 花样年华	32
	04	Riffle 涟漪	36
	05	The fabulous MasterKong 最靓的康师傅	40
	06	Shining shadows 彩影	44
	07	Shadows upon shadows 叠影重重	48
	08	City silhouette 城市魅影	52
	09	Firework 火树银花	56
	10	Floral wind-bell 花语风铃	60
	11	One day in spring 春野	64
	12	Sunshine in polar night 极夜的阳光	68
	13	Ever as moon, risen as sun 如月之恒，如日之升	72
	14	City of memory 忆想城	76
	15	Forest in the south pole 南极森林	80

16	Dancing with the wind 暗香浮动	84
17	The sounding cans we tread back in the days 那些年我们一起踩过的发声易拉罐	88
18	Mottled shadows of the trees 树影斑驳	92
19	Ice-like hexagon 六棱冰	96
20	Pandora's door 潘多拉之门	100
21	Ring 铃	104
22	9:30 PM 九点半	108
23	Blossoming 绽放	112
24	Light house 光之小屋	116
25	A midsummer night's dream 仲夏夜之梦	120
26	Life as flowers in summer 生如夏花	124
27	Spinning light 旋光	128
28	An impression on Zao Wou-ki 赵无极印象	132
Part 3	Students' comments 学生感言	136
Part 4	Site assessment 现场评审	140
Part 5	Review of years of teaching 历年教学回顾	144

Light and shadow, heart and shape

Ever since the existence of light, numerous shapes of shadows follow.
Ever since the existence of architecture, light and shadows accompany it.
As the Chinese sayings go: "Hou Yi shooting the suns", "Li Bai staring at the moon" – natural light gives buildings vivid lives; "Ying people holding the candle", "Watching lanterns on the Lantern Festival" – artificial light expands the horizon of cities and architecture. In the industrial society, "Edisons" lit up the night sky and every corner of buildings with electrical lights; while in the information society, new technologies emerge. LEDs, with less energy consumption and more flexibility, emit even more beautiful light and colors, leaving people soul-stirring light and shadow.
To help future architects understand the charm of light and shadow, Professor HAO Luoxi and her team have been teaching the course "Architecture Physics: Lighting Environment" for 15 years. This is a fundamental, 17-hour course for sophomores of College of Architecture and Urban Planning (CAUP), Tongji University. The course hour is short, however, both teachers and students exhibited great involvements, and very much enjoyed the wonderful teaching and studying processes. The three books presented here in front of us today, including "2015·Let's be Your Light", "2014·Light Shadow and Season" and "2013·The Light of The Antarctic", are the achievements from cooperation between teachers and students during the years 2013 – 2015.
We are moved by such achievements. These are teaching and experiments with rich characteristics of the time. They are combinations of art and technology, through cooperation between teachers and students.
They think and work: arrange light sources, design devices; they learn from history and raise questions to future, wandering around in the free realm of light and shadow, being intoxicated. They organize exhibition visits, design dramas, prepare music, and turn pure lighting technology into comprehensive art exhibitions as well as lighting festivals for teachers and students.
All of these stemmed from their desire to pursuit technology improvements, and their wish mood to seek various art forms. The forms to express vision and longing comes from their desire and wish.
We look forward to seeing Professor HAO and her colleagues to continue the innovations and "emit more gorgeous light". We appreciate people who offered their help during the series of teaching processes. We remember the warm shadows they left, and wish students great achievements in future study and creation, and "exhibit more beautiful colors".

Prof. Dr. LI Zhenyu
February 22, 2016 (The Lantern Festival)

光与影，心与形

自从有了光，便有了千姿百态的影。

自从有了建筑，光影便一直与之相伴。

后羿射日，李白对月，自然光赋予建筑有声有色的生命；郢人秉烛，元宵看灯，人工光拓展了城市和建筑的时空。进入工业社会，爱迪生们用电灯照亮了夜空，照亮了建筑的每一个角落；进入信息社会，新技术层出不穷，LED 以更少的能源，更便捷的组合，发出更加美妙的光色，留下更加动人的丽影。

为了让未来的建筑师们理解光影的魅力，郝洛西教授与她的团队开设"建筑物理（光环境）"已经有十五年了。这是一个 17 课时、面向同济大学建筑系本科生二年级的专业基础课程。课时虽少，但参加课程的老师、同学们却非常投入，非常享受这样美好的教学过程。今天呈现在大家面前的三本书《2015·流光魅影》《2014·光影时节》《2013·南极之光》，就是 2013-2015 年期间师生共同合作的成果。

这样的成果，让我们为之感动。这是富有时代特征的教学和实验，是艺术和技术的结合，是老师和学生的合作。

他们动脑动手，布置光源，设计装置，向历史学习，向未来提问，徜徉在光和影的自由王国里，如痴如醉；他们组织展览，策划剧情，配制音乐，把单纯的光的技术，变成了综合的艺术展示，变成了师生共同的光影节日。

这一切，都是因为他们有追求技术进取的心愿，追求艺术多态的心境。有了心，才有了表达情怀的形。

我们期待郝老师和她的同事们不断创新，发出更加绚烂的光；我们感谢给予这一系列教学活动各种帮助的人们，记住了大家留下的热情的影；我们祝愿同学们在未来的学习和创作中大有作为，呈现出更加美丽的光与色。

院长
2016 年 2 月 22 日元宵节改定

Inheritance and development

The book series on your hands belongs to the series of Tongji "Architecture Physics: Light Environment", which is the outcome of fifteen years' teaching exploration and innovation. This book, for the first time, attempts to introduce the format of digital media besides the paper-based publication. Readers can download the APP and have an intuitive experience of the review site, and fully appreciate the splendor of each work in the books. This series includes three books: the first book is "2015·Let's be Your Light" published in 2015; the second book is "2014·Light Shadow and Season" in 2014; the third one is "2013·The Light of The Antarctic" in 2013. Due to the limited space, teaching achievements before 2012 are not included in this publication; however, years of work still exudes the unique light, such as the large-scale installations "sound and light show" during the 2010 Shanghai World Expo; the work "bloom", after the demonstration in Expo cultural center, continues its applications in healthcare buildings, passing the infinite charm of light and space.

As an important fundamental course, "Architecture Physics: Light Environment" is designed for the sophomores with an architecture major. Given only 17 credit hours, the course is taught in three main aspects: (1) light, color and visual environment; (2) the light source and lamps selection (laboratory teaching); (3) quantity and quality of lighting. As there are courses "Interior Lighting Art" and "Daylight and Architecture" for senior students, this course can be seen as the entry-level courses for students to have a feeling of light and illumination. Over the years, the coursework topics are carefully designed by the teaching team, focusing on the light and optical properties of materials, shape, light-emitting device, the control concept, light drawings and expression, and other rich and comprehensive understanding. This helps students to have innovative thinking, practical hands-on operation, and the attitude of pursuing excellence in design works. As a team, students cooperate to complete the job. Everything is designed for students' self-learning, exploration, and practice.

Although the course is limited to six weeks and three credit hours per week, the students are sophomores who just stepped into the design world, and furthermore, the students must ensure that their main courses are not much affected, the students succeeded to finish the good work in the form of team work, with the limitation of defined theme, venue, time, number and color of LEDs, the conditions of environmental protection waste materials. The last day of teaching is for the show and assessment of homework. Therefore it can be treated as a time-limited quick architecture design, all the on-site dynamic changes are students' "hand control". The huge number of constraints are indeed challenges to both teaching and learning sides, and this is why I, as the teacher for this course, always be amazed by students' works.

We appreciate the support from the president of the Tongji University Press, Prof. ZHI Wenjun. He always pay close attention to the team's teaching efforts. Prof. ZHI have suggested us to publish our teaching efforts many times. Due to the fact that it is really difficult to reflect the content and expression of students' coursework by paper-based publications, this effort had been delayed to seeking the opportunity of digital-media publication. This summer, Prof. ZHI and the president of Jin Shang Digital, Mr. Xiaobo Xing, had a discussion with me at the Tongji University Press regarding the publication of this teaching effort. Ultimately, we decided to utilize both paper-based and digital-based formats to publish the three books, to share our experience and results from our 15 years of teaching on the course "Architecture Physics: Light Environment".

Special thanks to CAUP colleges, graduate and undergraduate students, and friends from industry who participated in this teaching work. Especial thanks to LIN Yi and CUI Zhe, for their contribution in curriculum construction.

As 2015 is the "International Year of Light" declared by the United Nations, to commemorate "the father of optics" - Ibn Hazm, who wrote an important optics book a thousand years ago, and a series of important finding in optics history, we would like to take the opportunity to pay tribute to the "International Year of Light"!

HAO Luoxi
12/15/2015

传承与开拓

您手中的这套书是同济大学"建筑物理（光环境）"这门课程 15 年教学探索与创新实践的成果汇编。这套书尝试在纸媒出版的基础上引入数媒，读者下载 APP 就可直观体验到作品评审时的现场情况，充分领略每个作品的光彩。本套书共计三册，第一册是 2015 年的作业汇编《2015·流光魅影》；第二册是 2014 年的作业汇编《2014·光影时节》；第三册是 2013 年的作业汇编《2013·南极之光》。由于出版篇幅所限，2012 年之前的教学成果没有列入此次的出版计划，但那些年的作品依然散发着特有的光芒，如 2010 年上海世博会期间的作业"声光 SHOW"大型装置；再如当年的作品"绽放"，历经世博文化中心的演绎应用，作为后世博的再应用，继续在医疗健康建筑中，传递着光与空间的无限魅力。

作为建筑学一门重要的专业基础课程，"建筑物理（光环境）"面对建筑学专业二年级的本科生授课，鉴于只有 17 个学时，主要讲授三个内容：光、颜色与视觉环境；光源与灯具选型（实验室授课）；照明的数量与质量。对四年级本科生开设有"室内照明艺术"和"日光与建筑"，因此可以看作是同学们接触光与照明知识的入门课程。历年课程作业选题来自教学团队的精心设计，重点在光、材质及光学特性、形态、发光器件、控制概念、光照图式及表达等丰富而综合的理解，培养学生开拓创新的思维能力、实践操作的动手能力、追求卓越的设计态度。作业完成以小组为单位，培养学生协同攻关的能力。一切向着自主学习型、设计探索型和实践应用型的学习方式迈进。

尽管只有为期六周、每周三个学时的授课，尽管他们只是初涉设计的二年级学生，但他们在保证专业课程设计作业不受影响的情况下，以团队合作的形式，在限定主题、场地、时间、LED 颗粒数量和颜色、环保废弃材料的条件下，完成最后的作业。教学的最后一天进行作业展示及评审。因此也可以说，这是一个类似建筑设计快题的限时作业，现场所有的动态变化均是学生们的"徒手控制"。如此多的限定条件，对教与学来说都是莫大的挑战，我作为任课教师和大家一样，对学生们最后呈现的作品总是每每发出由衷赞叹。

承蒙时任同济大学出版社社长的支文军教授一直以来对团队教学的关注，多次讨论将该教学的成果汇编成册出版，但苦于仅凭纸质的确很难全面反映该课程学生作品的内容与表达，于是就心有不甘地一直寻求数媒出版的可能性。今年暑假期间，支教授连同"今尚数字"邢小波先生约我到同济大学出版社，一起探讨出版事宜。最终我们决定以纸媒加数媒的出版形式，由张睿老师担任责任编辑，出版这三册书，分享 15 年来承担的"建筑物理（光环境）"课程的教学成果。

在此特别致谢团队中共同进行课程建设的林怡、崔哲两位老师，还有参与教学工作的学院老师、团队研究生、历届本科生及企业界朋友。

今年适逢联合国宣布 2015 年为"光和光基技术国际年（简称国际光年）"，以纪念"光学之父"伊本·海赛姆的光学著作诞生一千年及光科学历史上一系列重要发现，借这套书的出版，向国际光年致敬！

任课教师
2015 年 12 月 15 日

致南极中国长城站第29次科考队全体越冬队员

最长的夜，最靓的 光

建筑物理光环境——光影构成作品汇报展示
2011级建筑学专业

南极大陆与世隔绝，地理环境特殊，在众多领域有着重要的科考意义。近年来，越来越多的中国科考队员前往南极进行考察，但是南极单调的视觉环境和枯燥的极夜生活影响着科考队员的身心健康。针对南极的这一特点，作业要求学生制作适合极地站区的光艺术装置，以丰富视觉元素，改善室内光环境，营造温馨的生活氛围，陪伴南极越冬队员度过漫长的极夜，保证越冬队员的心理健康。

此次作业要求学生采用低碳环保的LED光源，结合南极站区日常生活中产生的可回收废弃材料，基于LED的发光特点和材料的光学特性，尝试光、材质及形态的创新设计。其中优秀作品将随中国南极第三十次科考队"雪龙号"船运往长城站，在最长的夜中为越冬队员点亮最靓的光。

作品展示：5月11日（周六）19:30~22:00
展示地点：建筑与城市规划学院ABC广场
顾问指导：中国极地研究中心
技术支持：上海亚明照明有限公司
任课教师：郝洛西 教授

祝词 Congratulations

  同济大学建筑学专业 2011 级的同学们，大家晚上好，我是中国第 29 次南极科学考察队长城站站长俞勇，获悉你们将于今晚举办"最长的夜，最靓的光"LED 灯光艺术装置最后的评审与展示活动，我在离你们一万六千五百多公里的南极大陆乔治王岛向你们问候，并预祝你们今晚取得成功。

  同学们，你们的任课老师郝洛西教授是我们长城站度夏队员。两个多月以前，她在这里已完成了极地站区光健康研究的现场实验工作，回到了你们中间。我想她一定向你们介绍了南极的气候环境特点以及站上的生活、工作情况。这里远离祖国，与世隔绝，视觉单调，对我们考察队员的身心是极大的考验。相信你们的作品一定会给我们带来丰富的视觉感受，以减轻极夜的单调和枯燥。

  同学们，我在南极注视着你们，更期待着你们的精彩作品来到长城站。祝你们成功，加油同学们，并希望你们享受今晚的活动。谢谢大家！（文字源于视频录像）

中国第 29 次南极科学考察队长城站站长

2013 年 5 月 11 日

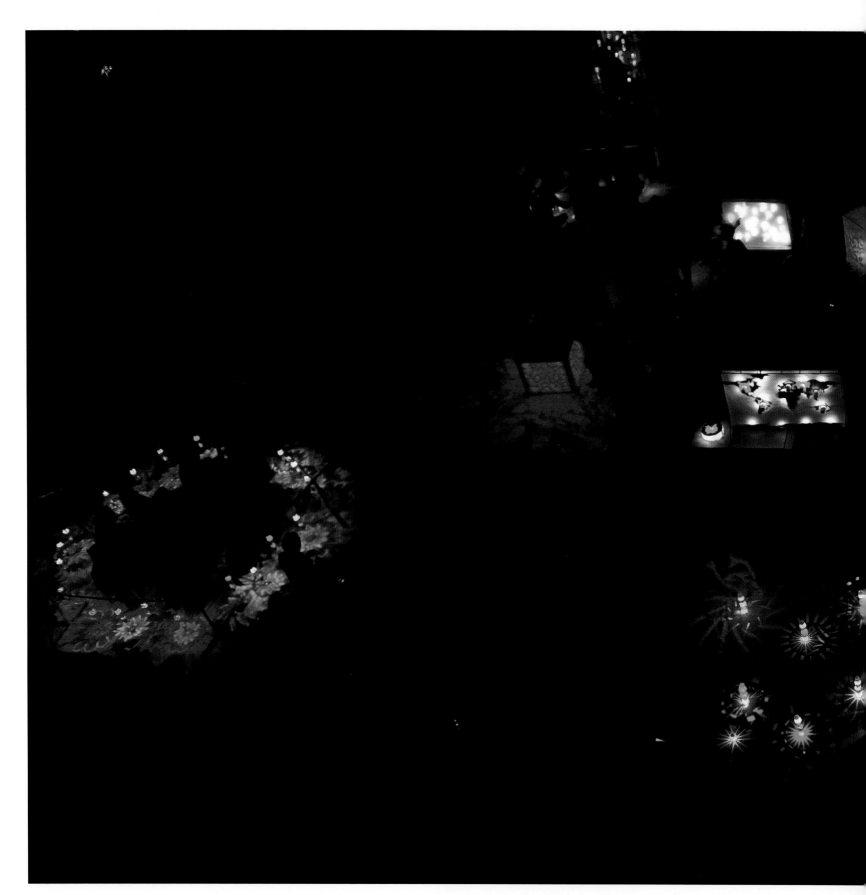

作业简介 Task introduction

一、作业题目

最长的夜，最靓的光

二、作业目的

通过对南极科考站区的了解，亲自动手制作光艺术装置，以改变南极站区的室内光环境，改善极夜时科考队员的情绪。

作为低碳环保的新光源，LED 是这次作业所采用的发光器件，结合生活中常见的材料，基于 LED 发光特点和材料的光学特性，尝试光、材质及形态的创新设计。在这个实验性教学过程中，发挥团队精神，努力探索，通过焊接 LED 的操作实践以及光装置的创意设计，观察各种材料、光线及环境的互动关系，发挥大胆和丰富的想象，用光描绘出一幅幅绚烂的艺术构成，呈现出一场独特的视觉盛宴。

三、作业内容与方法

1. 使用材料

材料不限。推荐使用生活过程中产生的废物，如啤酒瓶、易拉罐、矿泉水瓶、方便面袋子、垃圾袋等，循环利用，力求对环境的影响的最小化。

2. 光源选择

LED 芯片，共有 10 种不同光色，每组数量统一，不同光色各 10 颗，共 100 颗。

3. 方法

LED 焊接：利用电烙铁等工具将 LED 芯片与电线焊接。

装置设计：探索光与材料的特性，进行光艺术效果实验，选择符合主题的效果。

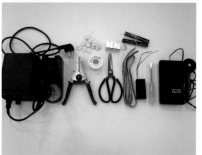

四、作业组织

以班为单位分成工作小组，每组 5-6 名学生。

五、作业成绩评定

按照工作小组进行公开评审，每小组一个成绩。优秀作品将随中国南极第三十次科考队"雪龙号"船运往长城站。

六、作品展示

展示时间：2013 年 5 月 11 号（周六）19:00-22:00
展示地点：建筑与城市规划学院 ABC 广场
评审颁奖：文远楼 215 室 21:00-22:00
顾问指导：中国极地研究中心
技术支持：上海亚明照明有限公司

南极之窗，炫彩立方
Antarctic window, colorful cube

作者 Designers

李方芳　刘子晨　刘茜　金勋　路加

概念原理 Concept

南极的环境是极端恶劣的，建筑成了这里唯一的庇护，而窗户却丧失了原来的功能。因为在这里，人们不再渴望透过窗户了解外面的世界——单调的冰天雪地。我们装置设计的核心是希望回归窗户的本质，给人一种外眺欲望，形成一道别样的窗间风景。我们对传统"冰裂纹"花窗进行演绎，在立方框架内构造多层花窗板，并可以使其固定在窗户上。白天自然光为光源，夜晚LED灯为光源，通过全天候的光影变化，重新赋予窗户以丰富多变的视觉享受。

材料构成 Material

设计草图 Sketches

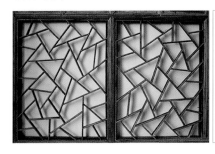 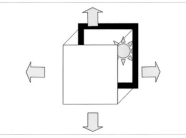   

制作过程 Construction

  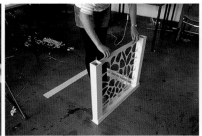 

成果表现 Realization

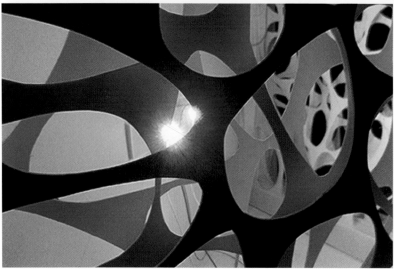
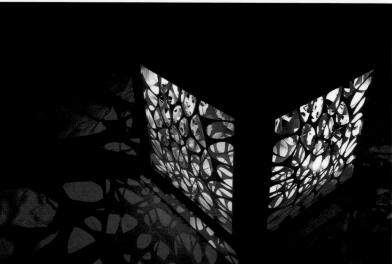

咣咣咣光
Guang guang guang light

作者 Designers

王建桥　傅宸彬　张弛　张松岳　韩丞奎　金春旭

概念原理 Concept

我们最初的概念就是将光与音乐结合起来，想到利用音响播放音乐时产生的震动，使光线随音乐产生变化。将细木条一头连上泡沫球接触音响，另一头粘上带有大小圆孔的纸片。LED灯放置在一圈六个音响中间，灯光透过纸片向外照射。在外侧的立方体纸屏上便显现出颜色大小各异的随音乐跳动的光斑。同时，采用单片机来控制部分LED灯，使它们按照音乐的节奏来开关，产生音乐控制光的闪烁效果。整个装置不仅很好地将灯光与音乐结合起来，还实现了自动控制。

材料构成 Material

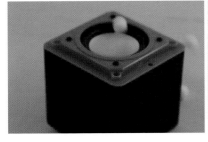  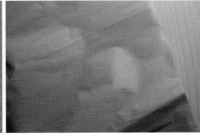 

设计草图 Sketches

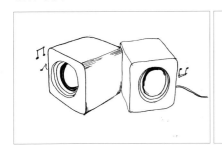 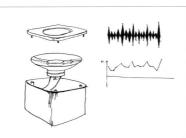 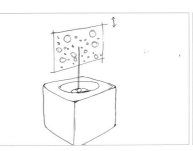 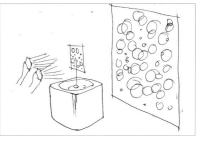

制作过程 Construction

    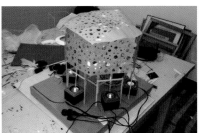

成果表现 Realization

花样年华
Golden days

作者 Designers

傅 荣　王瑞琦　周瑾楚　章 晓　彭程瀚　单浩然

概念原理 Concept

花的美丽与万花筒的绚丽都能给人愉悦，而光的加入使得效果更好地得以体现。南极——一片白茫茫的世界，颜色绚丽的花朵自是难以一见。我们的装置，可作为灯具的同时，在旋转时更具有如万花筒般绚丽的效果，可以带给队员们在当地自然环境中难得的愉快感受。

材料构成 Material

 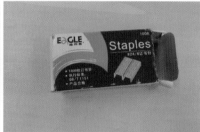  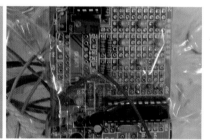

设计草图 Sketches

  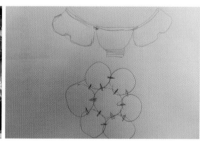 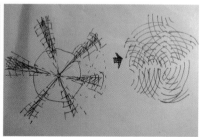

制作过程 Construction

成果表现 Realization

涟漪
Riffle

作者 Designers

概念原理 Concept

长期生活在南极的人，接触最多的就是冰和水。然而残酷的环境使人与水变得敌对。利用光，我们让人与水形成生动的互动。我们利用日常的生活垃圾——塑料可乐瓶的中部，在一侧放灯将光打到承影面上。瓶身原有的圈在光的照射下自然形成水纹，而在中部汇聚的奇妙光影形象地模拟了水。通过改变LED灯与瓶身的距离，形成水波散开的涟漪效果。

材料构成 Material

设计草图 Sketches

 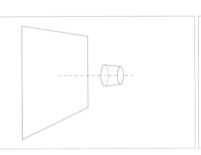 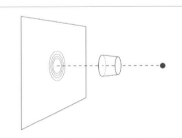 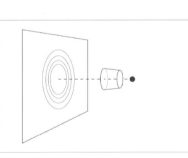

制作过程 Construction

   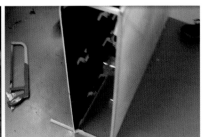

成果表现 Realization

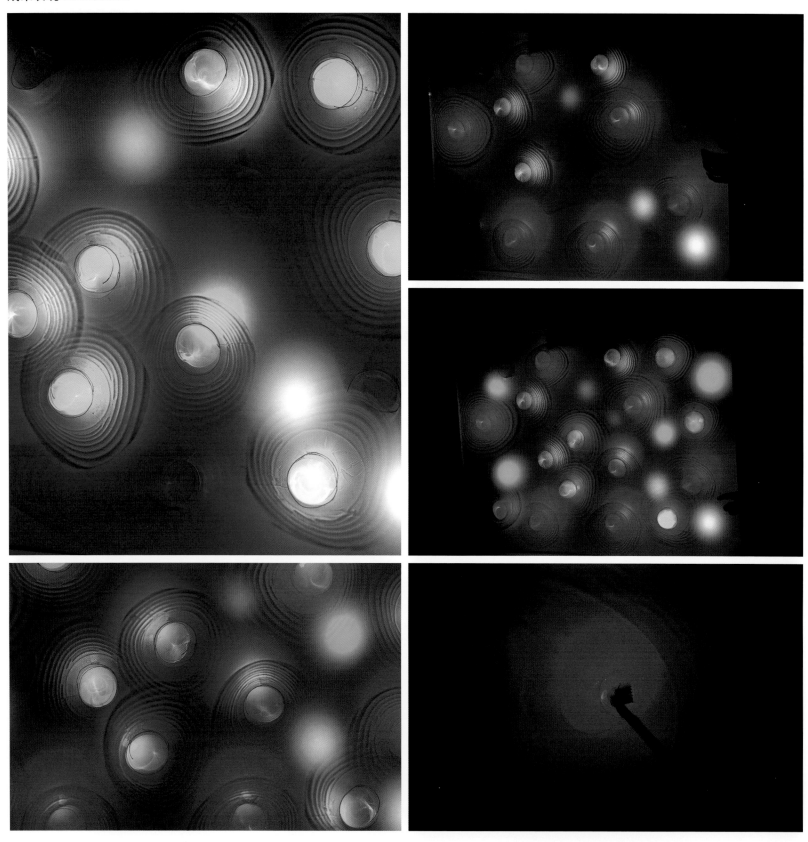

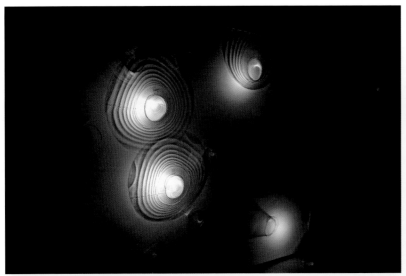

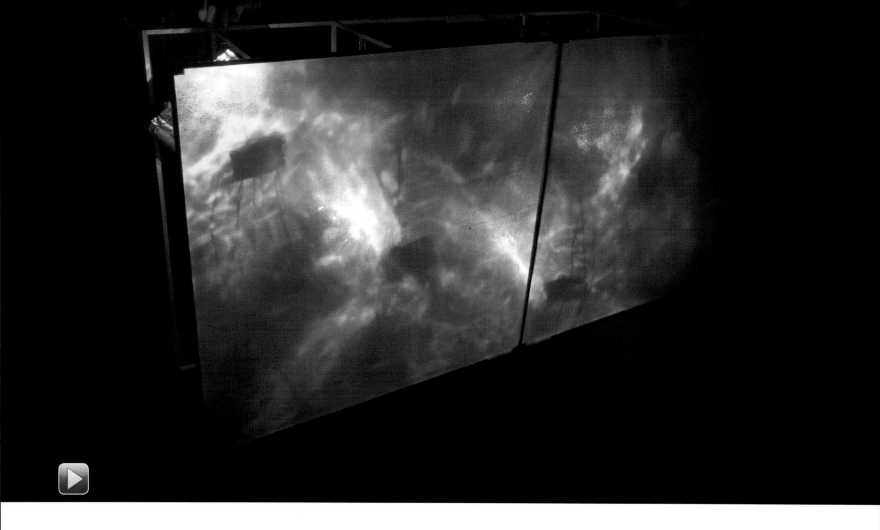

最靓的康师傅
The fabulous MasterKong

作者 Designers

彭书勉　何凌芳　潘逸瀚　孙一桐　奚祎婷　闵　睿

概念原理 Concept

在南极，方便面是常备食物，但是方便面的袋子却是废品。因此我们变废为宝，利用即食方便面袋子内部的光学特性，将灯放置在袋子反光面之前，在承影面上形成丰富的效果，呈现如同星云，又如同极光般的梦幻光影。再将其与转轴与架子相结合，可以将袋子旋转，推进，在光与人的互动中产生更有趣的效果。

材料构成 Material

  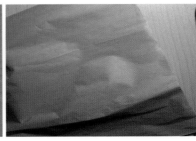 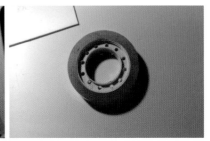

设计草图 Sketches

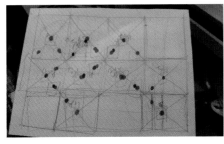 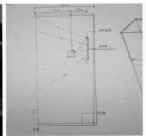 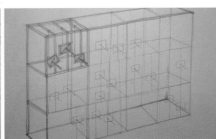 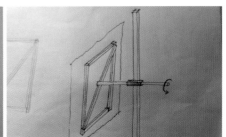

制作过程 Construction

成果表现 Realization

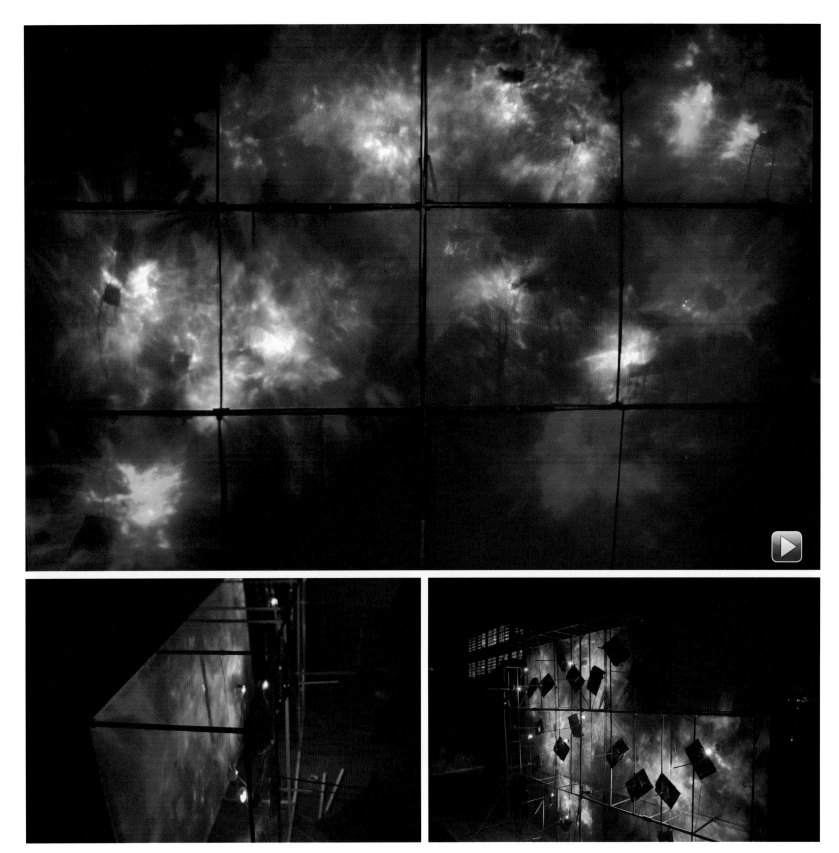

彩影
Shining shadows

作者 Designers

张润泽

洪日环

李在丞

权恩真

李彦柱

尹恺宇

概念原理 Concept

光与影是对立的，更是相辅相成的，通过对影子的研究，我们发现在单光源照射形成的影子上再打上其他颜色的光就可以让影子出现不同的色彩。一个光源就可以形成一个影，那么将三色光通过阵列的形式排开，再根据上述原理就可以照射出多重炫彩的影子，而遮挡物体（可以是人或物）在不同的位置就会形成不同形状的影子，从而既保证了装置的活动性又达到了色彩丰富美丽的效果，给极地的黑夜送去炫丽的色彩。为了环保，我们还废物利用，用瓦楞纸箱做装置主要材料。

材料构成 Material

设计草图 Sketches

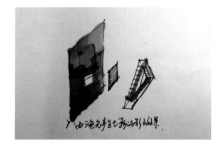
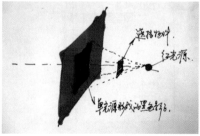
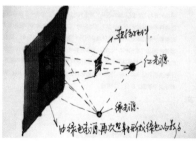

制作过程 Construction

成果表现 Realization

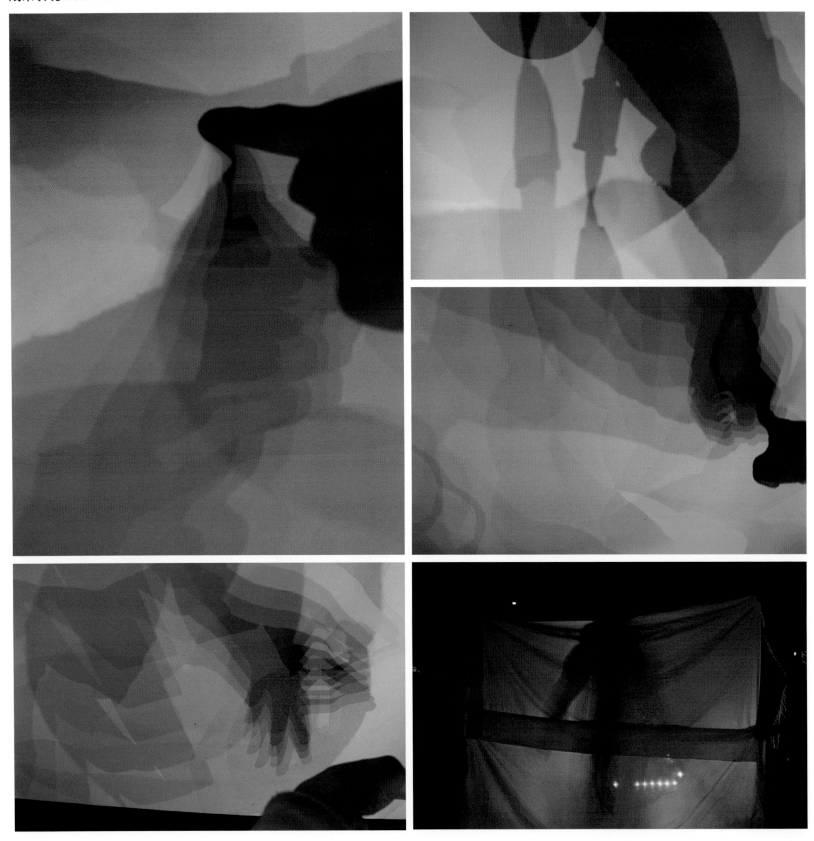

叠影重重
Shadows upon shadows

作者 Designers

李 冲　　罗君临　　陶依依　　周 姝　　欧 龙

概念原理 Concept

万花筒是我们儿时最爱的玩具之一，而它吸引我们的不只是其中美丽的图案，更是镜面反射出的图案无穷无尽的变化。因此我们将灯光置入其中，让无尽的反射带来绚烂的叠影，希望用色彩变化的万千重影给南极单调的景色带来别样的美丽。

材料构成 Material

设计草图 Sketches

  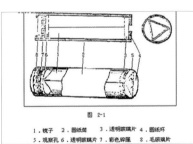 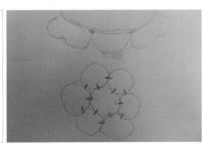

制作过程 Construction

成果表现 Realization

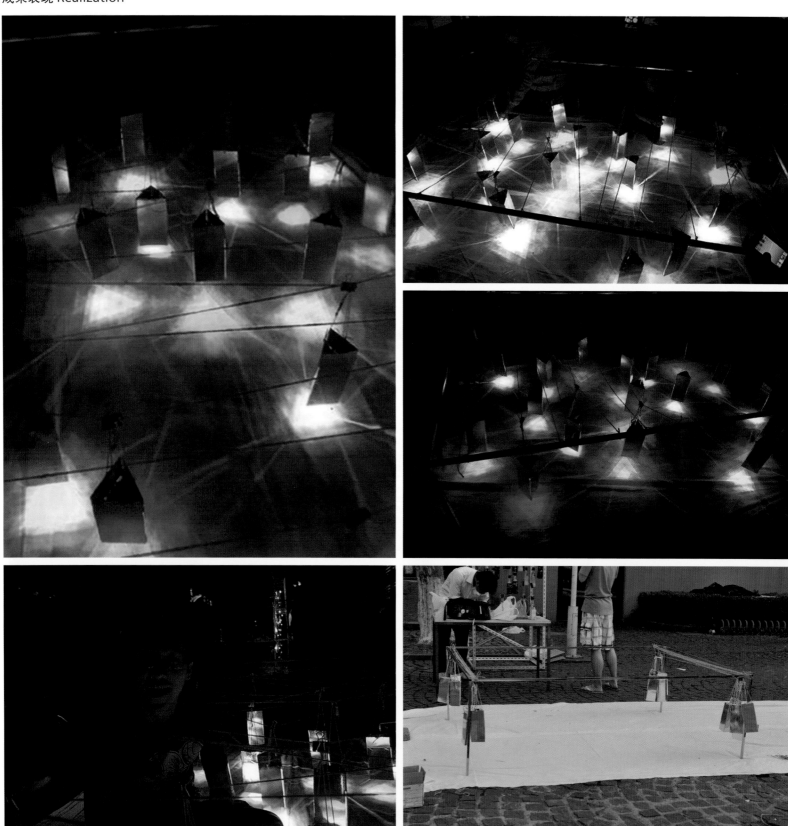

城市魅影
City silhouette

作者 Designers

张雨川　史 好　周瑞烽　顾云迪　何 进

概念原理 Concept

我们的概念源于城市剪影,考虑到极地工作人员长期远离城市的心情,所以利用灯光,还原城市生活,在漫长的极夜中给他们送去来自远方的问候。我们找寻有特色的城市剪影,如上海的东方明珠、澳大利亚的悉尼歌剧院,将它们重现到纸杯上,内部放置LED灯,在杯口安装转盘,最后将装置安装在用瓦楞纸板制作的世界地图上,用灯光模拟海洋效果的同时,"陆地"上的纸杯可以随着音乐旋转,纸杯内部的LED灯,将城市剪影投在地图上,产生奇特的效果。

材料构成 Material

   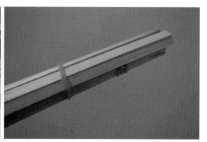

设计草图 Sketches

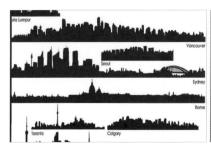 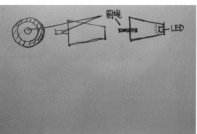 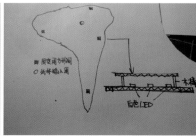 

制作过程 Construction

   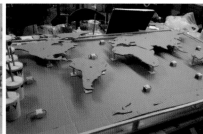  

成果表现 Realization

火树银花
Firework

作者 Designers

杜超瑜　朱忆濛　林哲涵　刘哲圣　李辰

概念原理 Concept

南极的黑夜是单调而漫长的，在节日里，对家的思念是南极科考队员最自然不过的事。我们希望用光模仿烟花绚烂绽放的效果，通过南极科考队员和装置的互动，让他们能够按照自己的意愿创造丰富的光效。用具有反光特性的包装袋，卷成筒形，并刻上长长短短的缝隙，从内部打出光线便可在承影面上形成长短不一、层次分明的线形光斑。装置内部灯光和筒的转动，增加了趣味。

材料构成 Material

设计草图 Sketches

  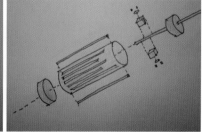 

制作过程 Construction

成果表现 Realization

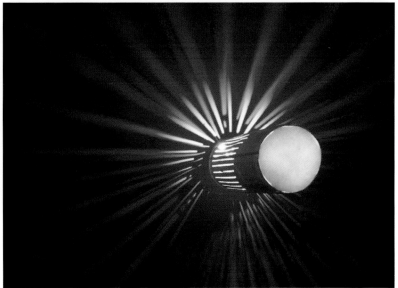

花语风铃
Floral wind-bell

作者 Designers

叶 帅　　叶心成　　柳兰萱　　赵静蓉　　达尔汉

概念原理 Concept

光的迷人之处，就在于它的随处而在、飘忽不定、万千色彩、变幻莫测。这几种属性令我们联想到随风飘动的风铃与大自然中艳丽多姿的花朵。希望通过装置，将两种美丽事物的结合，为人带来丰富、变化的视觉与空间体验。

材料构成 Material

  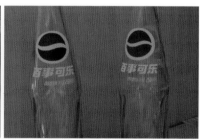 

设计草图 Sketches

 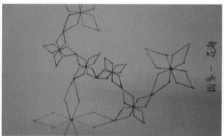  

制作过程 Construction

成果表现 Realization

春野
One day in spring

作者 Designers

沈彬　　吴迪　　刘欣鹏　　夏杨　　朴俊庆

概念原理 Concept

我们设计的意图是制作一个光互动装置，让装置和南极科考员发生有效互动。装置本身非常简单，只是由几片有机玻璃和一块泡沫板组成，但是当光打在有机玻璃的薄薄侧面上后，经过一系列复杂的折射反射，会形成梦幻般的光束。有机玻璃稍微有一微小变化，反射折射情况立刻变得复杂，投射出来的光线也极具梦幻色彩。配上班得瑞的美妙音乐《春野》（*One Day in Spring*），光线也跳起舞来。

材料构成 Material

 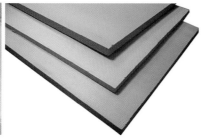  

设计草图 Sketches

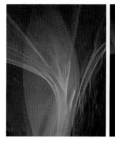  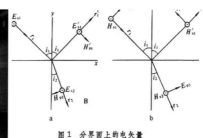 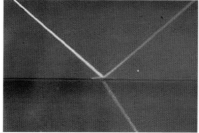 

制作过程 Construction

成果表现 Realization

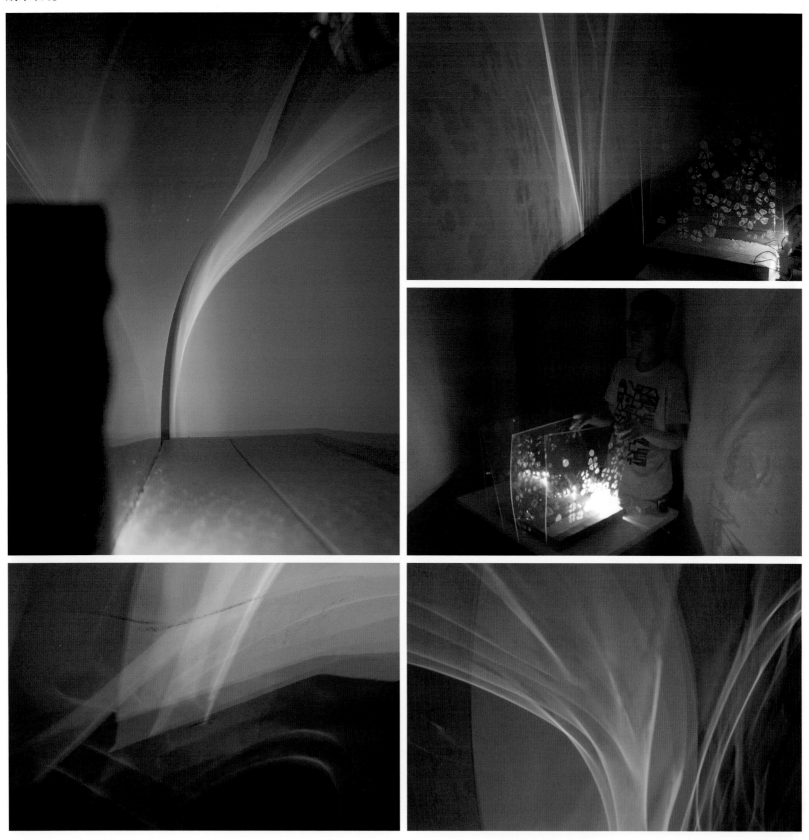

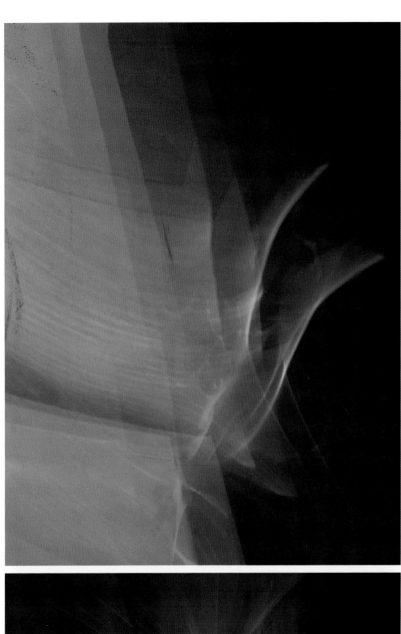
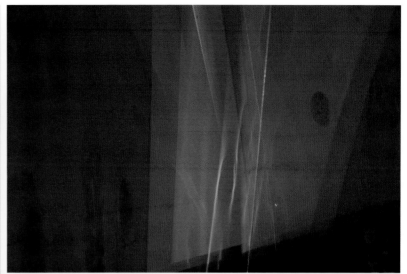

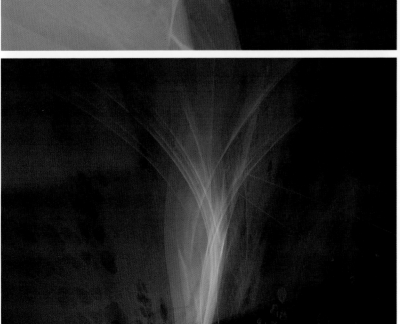

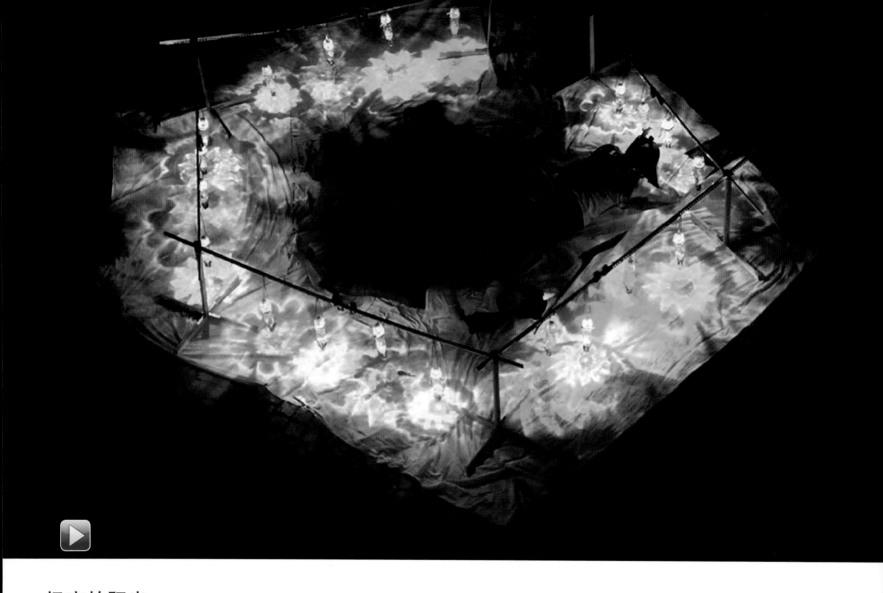

极夜的阳光
Sunshine in polar night

作者 Designers

马曼·哈山　张黎嫱　薛洁楠　王唯渊　马潇潇　张辰　马敏洁　张晗婧

概念原理 Concept

得知科考队员们将面临极夜，我们有了"创造太阳"的想法。受到一个玻璃杯形成的图案启发，我们采用了有螺旋竖向条纹的汽水瓶作为主要的装置材料。运用材料对光的透射、折射等特性，在地面上呈现太阳般的花纹，并配合声效，创造温馨的艺术效果。队员们被"太阳"环绕，在其中活动。愿这个装置能让科考队员在极夜中获得来自太阳的温暖。

材料构成 Material

设计草图 Sketches

 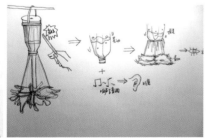  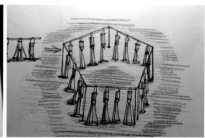

制作过程 Construction

   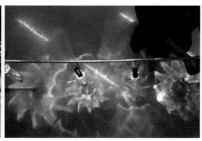 

成果表现 Realization

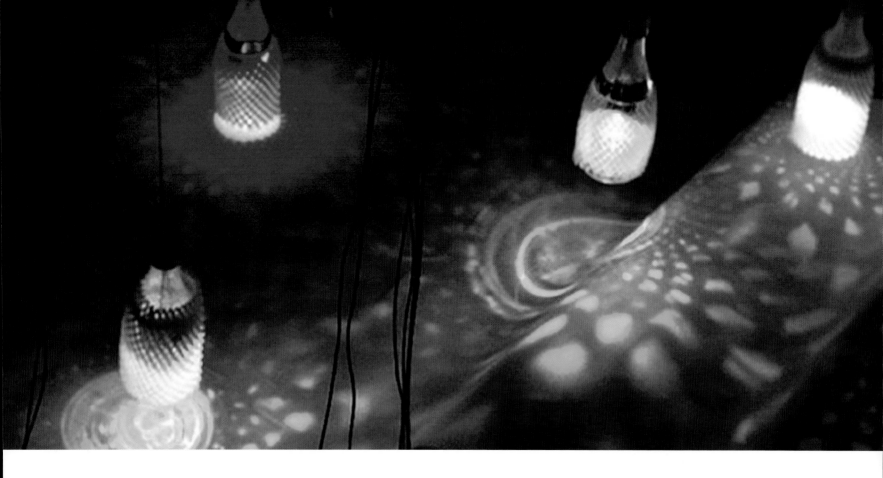

如月之恒 , 如日之升
Ever as moon, risen as sun

作者 Designers

王俊超　杨溪　黄蓟霖　张正秋　喻琳楠　任贝蕾

概念原理 Concept

果汁瓶瓶身的花纹与水果网套的格栅相映成趣，投射出花样绚丽的效果。四散流溢，恰如波光。瓶身内蕴含的光彩由冷入暖，简单意蕴月落日升的永恒规律。全部绽放时冷暖相就，日月共生。南极冷寒，极夜之时鲜见日光。星点光照期求弥补不足。装置一概用细绳悬于棚架之间，走过、路过、碰过、荡过，更添灵动绚逸。步行之间低头可见波光粼粼，犹如置身水底，更为陆上之民增添不可多得的乐趣。

材料构成 Material

 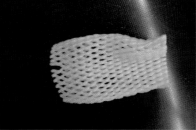  

设计草图 Sketches

 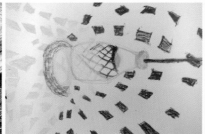  

制作过程 Construction

   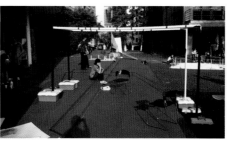 

成果表现 Realization

忆想城
City of memory

作者 Designers

吕德轩　刘雨涵　洪 菲　武晓宇　汪明澈　林笑涵

概念原理 Concept

尽管生活在城市里的我们常常会厌恶这座水泥森林的冰冷无情，但是，如真的走进极地，应该会怀念城市里的那些人、那些事吧。希望通过这样一个三层圆筒的光装置还原坐在都市行驶的汽车上向外看时的场景，看到远远近近的楼房快快慢慢、虚虚实实地从眼前飞过，从清晨到夜晚，思绪一下被拉回到那些记录着生活点滴的城市记忆里。

材料构成 Material

设计草图 Sketches

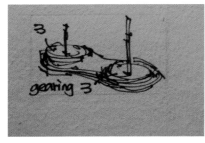 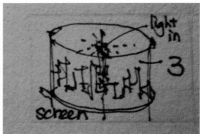 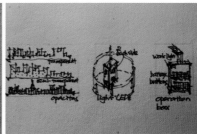  

制作过程 Construction

成果表现 Realization

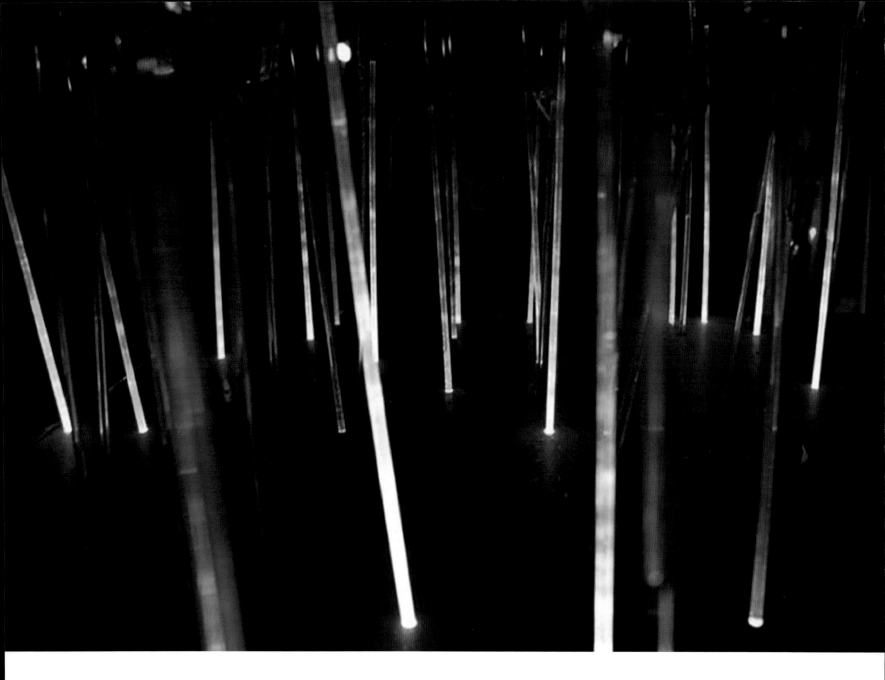

南极森林
Forest in the south pole

作者 Designers

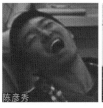     

陈彦秀　刘思雨　王余丰　严康妮　杨婉琦　赵世攀

概念原理 Concept

我们以南极的等高线和科考站的位置作为设计来源，希望做出一个点亮南极的荧光森林的灯光装置设计。装置分为两部分，第一部分作为背景灯，固定在支座上，灯光由下向上照亮导光棒。第二部分为导管，下部分连接灌着饱和浓盐水的气球，挤压气球的时候可以使盐水向上，最终接触导管顶端的导线，使顶端悬挂的灯点亮，让南极森林亮起来。

材料构成 Material

设计草图 Sketches

  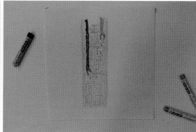    

制作过程 Construction

成果表现 Realization

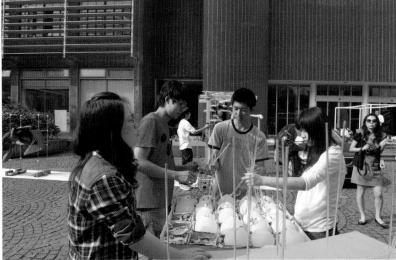

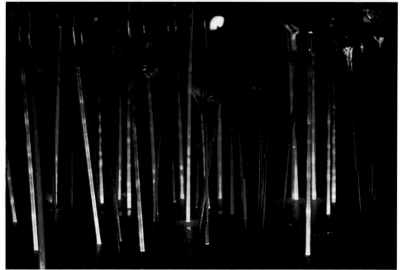

暗香浮动
Dancing with the wind

作者 Designers

李泽辉　杨珺卿　杨 萌　廖嘉文　王诗琪　江孟繁

概念原理 Concept

我们的概念源于中国古典园林的漏窗。首先用雪弗板切割出窗棂的图案，然后再用风车的旋转模拟窗外浮动的花影。当微风吹过时，风车转动，就营造出了暗香浮动的氛围。同时，用灯光模拟四季的场景，希望用带有浓浓的中国风的装置点亮南极的夜。

材料构成 Material

设计草图 Sketches

 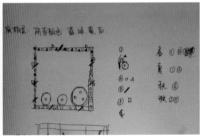 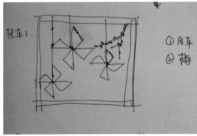 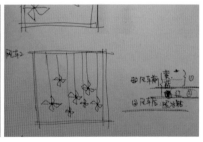

制作过程 Construction

成果表现 Realization

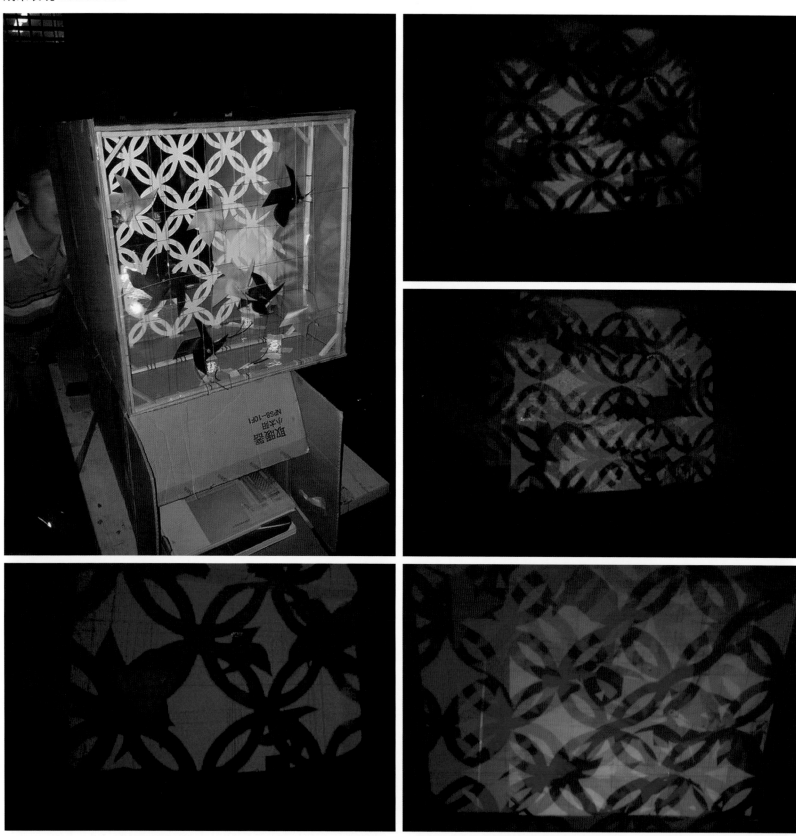

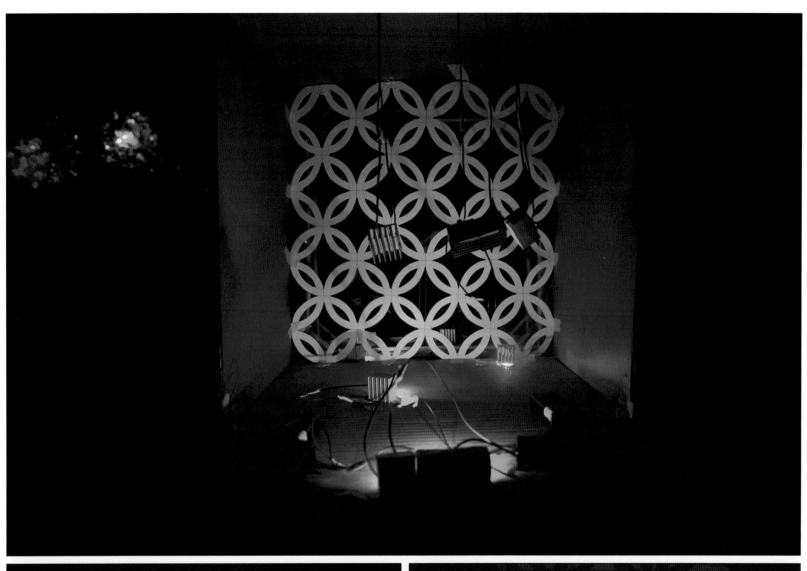

那些年我们一起踩过的发声易拉罐
The sounding cans we tread back in the days

作者 Designers

沙千陌　承晓宇　常琬悦　罗富缤　叶天娇　刘鹤　傅雷格　郑圭东

概念原理 Concept

极光是出现于星球高磁纬地区上空的一种奇特的发光现象。我们希望将这种绚丽多彩带入室内，为越冬的孤寂夜晚增添一抹靓丽的色彩。装置主材料为环保的废弃材料易拉罐，利用了易拉罐自身反光有弹性等特点，仅对材料本身进行简单的切割，将LED灯从顶部放入，罐内系的铃铛在绚丽的灯光下随风摇摆。队员们还可以通过踩压易拉罐形成声光互动的乐趣。

材料构成 Material

   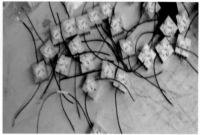

设计草图 Sketches

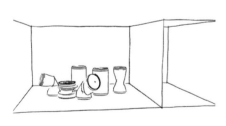 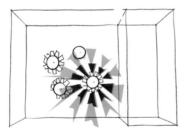 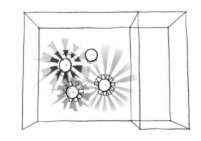  

制作过程 Construction

成果表现 Realization

树影斑驳
Mottled shadows of the trees

作者 Designers

方荣靖　孙宏楠　鲍芳汀　吕易达　杨柳青青　王晶晶　文森加

概念原理 Concept

在了无生机的南极大陆上很难看到绿色的影子，我们的这个装置让身处南极的队员们能够感觉到生机勃勃的大自然的存在。主要利用普通的碎纸屑，做了几个玻璃盒来充当回收废纸屑的垃圾桶，然后LED灯光穿过废纸屑，形成树影斑驳的效果。

材料构成 Material

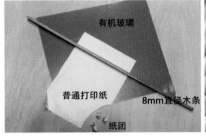   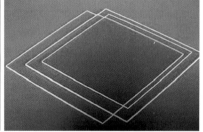

设计草图 Sketches

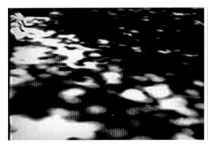   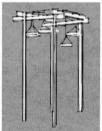 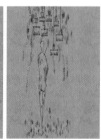

制作过程 Construction

成果表现 Realization

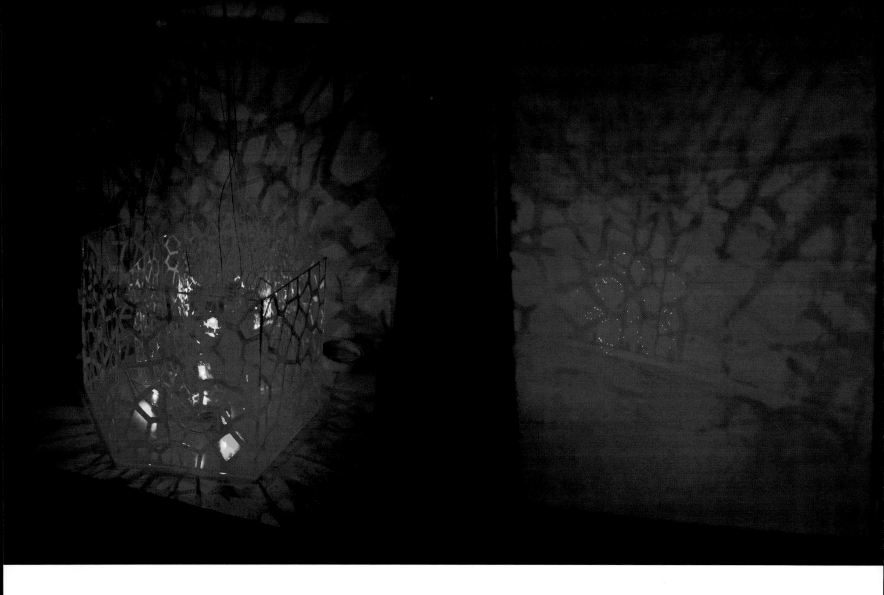

六棱冰
Ice-like hexagon

作者 Designers

吴剑翎　糜慕容　戴宜枫　邝远霄　董新基　黄成业

概念原理 Concept

南极多冰雪，所以我们在考虑能否通过一种装置来体现冰这种分子晶体的光学特性，希望通过这种特性来实现一种璀璨、晶莹、缤纷、错杂的光影效果，为南极科考站的漫漫长夜增添一抹靓色。通过查阅资料，我们了解到了冰的微观结构——由氢键作用于相邻水分子而组成六方晶系，即基本单元是六棱柱的分子晶体——于是我们确定了装置的六棱柱结构。为了丰富效果，我们设置了内、中、外三层不同大小的六棱柱并使之相对旋转，以产生更为缤纷炫目的效果。

材料构成 Material

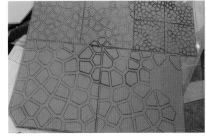  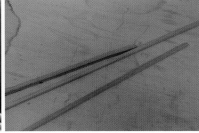 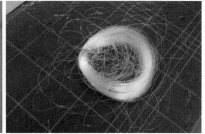

设计草图 Sketches

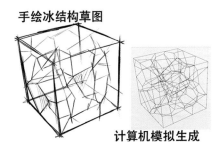 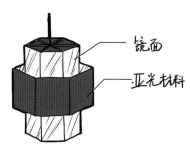 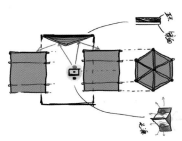 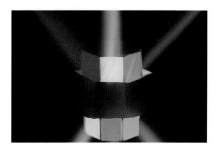

手绘冰结构草图　　计算机模拟生成

制作过程 Construction

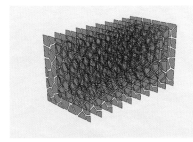 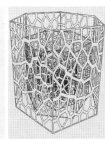 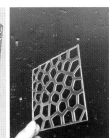  

成果表现 Realization

潘多拉之门
Pandora's door

作者 Designers

潘亦欣　王　谦　郑星骅　付　涵　刘昕雯　邓可田

概念原理 Concept

"潘多拉之门"的设计初衷是使用灯具做出一个可以穿过的门帘。装置使用的材料有书页夹、磨砂玻璃纸、搭配LED灯组成的灯管,以及组装式衣柜做成的框架(门框)。其中除了玻璃纸外,全部材料都可在使用后重复利用,不会额外制造垃圾。

表演时遮住门框的布缓缓掀起,在五彩的门帘后出人意料地走出一个人。这切合了作品的名字"潘多拉之门",潘多拉之门的开启象征打开未知的宝库,这正与南极的神秘气氛相吻合。

材料构成 Material

   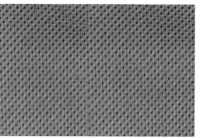

设计草图 Sketches

  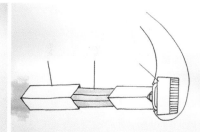 

制作过程 Construction

成果表现 Realization

铃
Ring

作者 Designers

韦寒雪　唐 韵　戴乔奇　汪 桐　王旻烨

概念原理 Concept

我们的装置将光线的多重散射和声音感知相互结合。选用的材料是在南极也多见的啤酒玻璃瓶和烧瓶。喝完酒后将啤酒瓶敲碎，放到烧瓶里，在酒瓶碎片当中放入 LED，不同颜色的酒瓶渣（琥珀色、绿色、透明、磨砂）搭配不同颜色的LED（红黄蓝白），在每个烧瓶中的LED可以一个、两个、三个互组合。圆形的烧瓶，使光线通过玻璃渣形成多重光线和丰富的光斑。通过旋转烧瓶，不同颜色的光斑也随之旋转，十分绚丽。

材料构成 Material

设计草图 Sketches

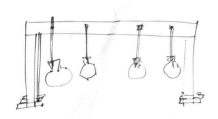  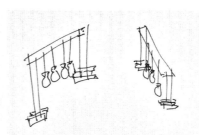 

制作过程 Construction

 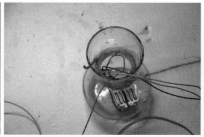  

成果表现 Realization

九点半
9:30 PM

作者 Designers

毛宇俊　严 轲　王 轶　王雨林　谢 超

概念原理 Concept

设计从南极的场地出发，考虑到在南极这样的一个环境中有许许多多的冰，希望利用冰来作为光线发散的介质。装置分为上下两部分，上部利用丝袜和亚克力鱼缸石（作为装置中冰的替代物）做成一个水滴的形态，并在其中放置一定数量的LED灯。下部用雪弗板、海绵和有机玻璃做成一个盒子，并在表面撒上一定的鱼缸石，和水产生一个水灯的效果。当上部装置中的冰慢慢融化后，落下的水滴在水灯表面产生涟漪的效果，并在两边的承影面上产生梦幻的光影效果。

材料构成 Material

设计草图 Sketches

  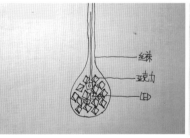 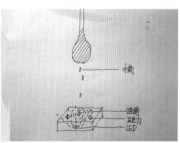

制作过程 Construction

成果表现 Realization

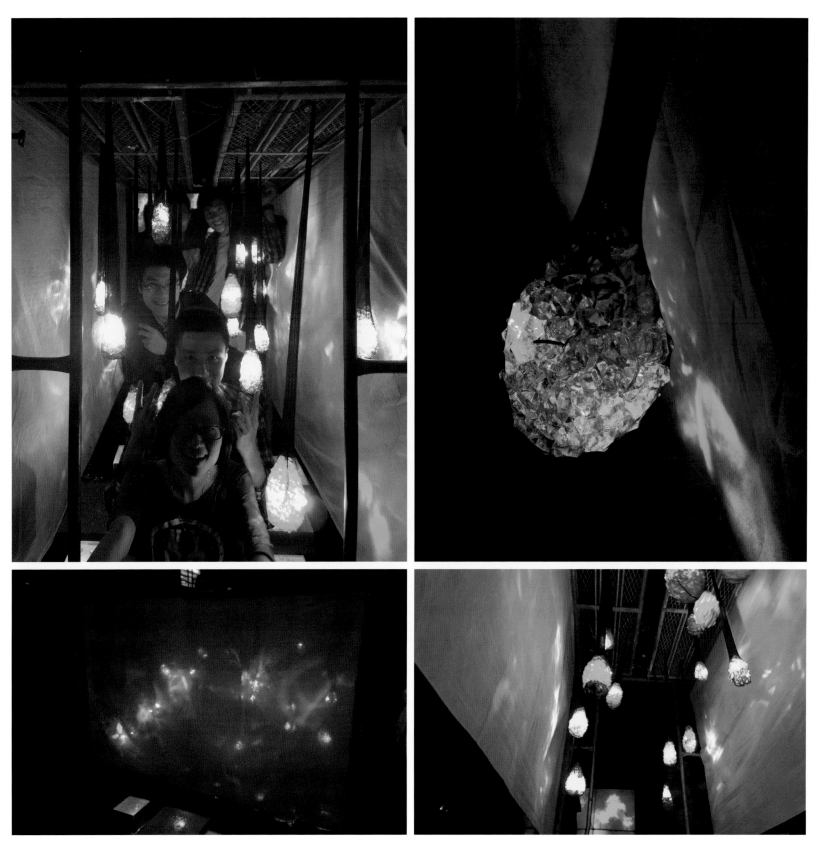

绽放
Blossoming

作者 Designers

曹诗敏　胡佳林　宋 睿　韩雪松　夏孔深　黄艺杰

概念原理 Concept

一个简单的矿泉水瓶在变化的灯光下会有很棒的光影效果。就用这些瓶装水来做一堵光艺术的隔墙，既可用来分隔空间，储存饮品，又能产生影响整个室内的艺术效果。因此我们设计了一个储水的架子，前置一块幕布，后挂一面装了 LED 灯的帘子，装置可大可小，可按实地需求进行设计。当人们掀开帘子取水的时候，幕布就会呈现绚烂的光影效果。

一些特定的节日、生日，还可以通过控制灯光的明暗色彩在幕布上打出各式庆祝图案，如字幕、爱心等。最终所呈现的光影效果是多样而不确定的，我们把这种可能性交给南极站的工作人员，让他们不仅享受光艺术的美感，更享受设计这一过程的乐趣。

材料构成 Material

设计草图 Sketches

  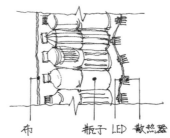  

制作过程 Construction

成果表现 Realization

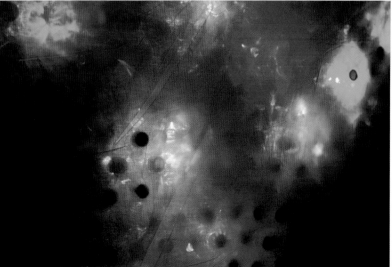

光之小屋
Light house

作者 Designers

吴潇　伍雨禾　蔡宣皓　张灏宸　刘庆　陈迪佳

概念原理 Concept

我们基本操作方法是对瓦楞纸箱再利用，光线照于其截面的孔洞之上时会出现极有趣的光影效果，同时希望能在大空间中营造出一个有趣的小空间，使科考队员们能在空余时光于小屋或是在此与他人进行互动，小屋中的空间就如同一个梦幻世界，处于屋内的人身上、墙上、屋上均是色彩斑斓、美妙无比，具有极强的包容性与代入感。

材料构成 Material

设计草图 Sketches

 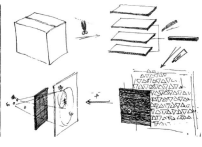 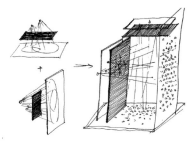 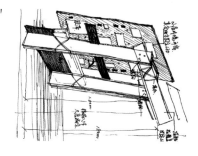

制作过程 Construction

成果表现 Realization

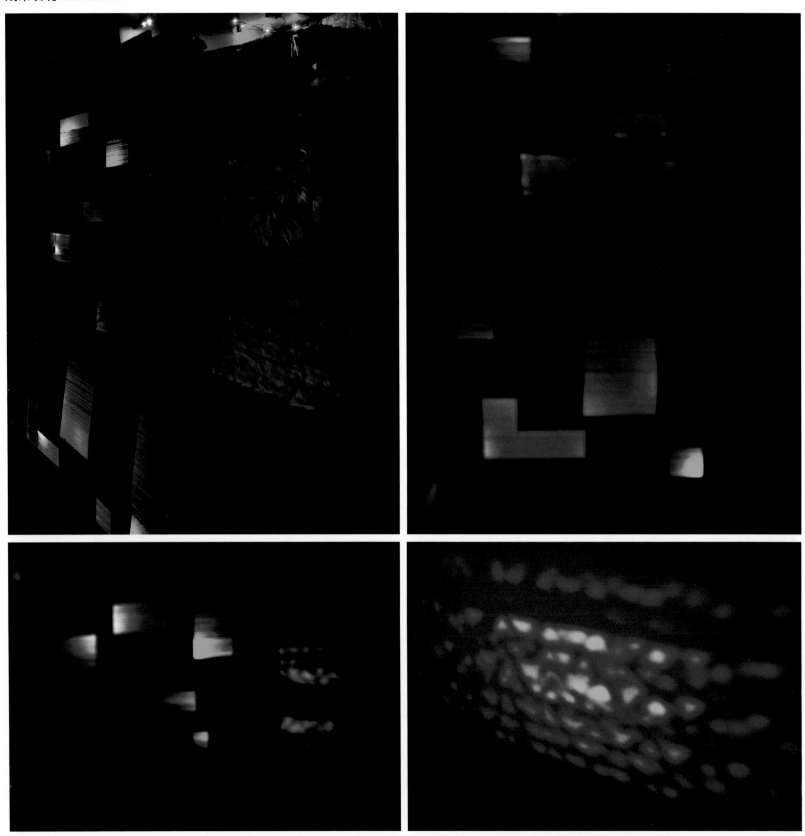

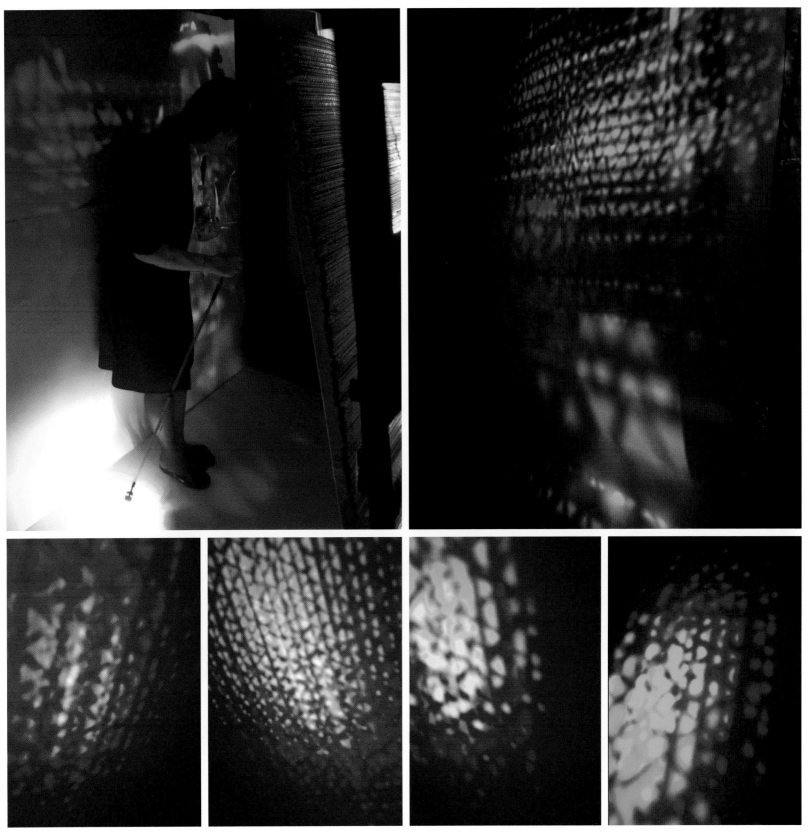

仲夏夜之梦
A midsummer night's dream

作者 Designers

钱静　黄芫　顾倩　刘一萱　姜晗笑　范雅婷

概念原理 Concept

灵感来自梵高的《向日葵》，绚丽绽放的颜色让人感受到了夏天的热情和梦幻，希望给在南极度过极夜的科考队员们带来仲夏的气息。运用易拉罐，在罐壁上刻出螺旋排列的一组弧形曲线，将弧形翻起，会形成一个引导光线的页片。同时，光线在铝制的内壁上形成反射，最终投射在墙上的效果是直射光和反射光的螺旋形组合。将易拉罐固定在墙上，通过不同颜色的灯光组合，可以拼贴出一面墙的仲夏。

材料构成 Material

设计草图 Sketches

 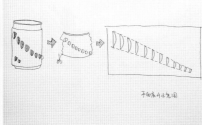 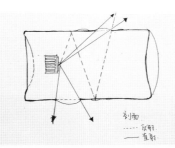 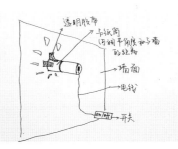 

制作过程 Construction

成果表现 Realization

生如夏花
Life as flowers in summer

作者 Designers

王 祥　张家宁　吕 宇　潘清伟　乔诚文　李慧妮　白承烨

概念原理 Concept

"生如夏花"选自泰戈尔诗集的一个标题，代表生命的璀璨。我们组的光装置欲借此意在初夏展示光彩的多重绚丽幻影。光线透过纸杯上的各种韵律镂洞会在地面上产生美丽的图案，我们再次用多个开洞的杯子叠加垒成塔状，则可以在地面产生多重梦幻光影，绚丽夺目。

材料构成 Material

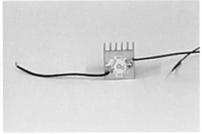

设计草图 Sketches

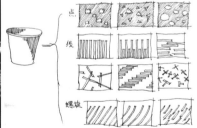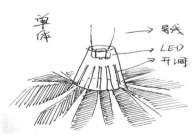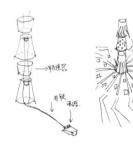

制作过程 Construction

成果表现 Realization

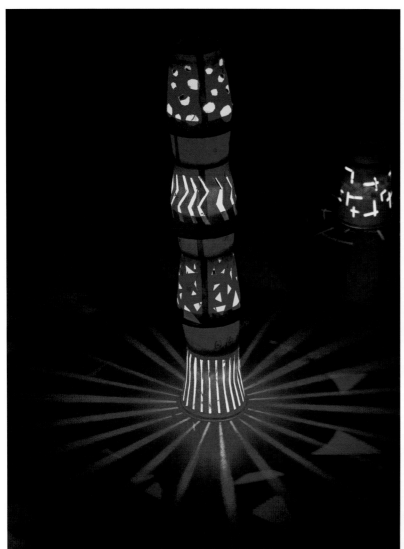

旋光
Spinning light

作者 Designers

程静瑶　　丁一　　郭芸颖　　高琳婕　　龚运城　　何蕾丝　　梁 宇

概念原理 Concept

南极的极夜给我们的印象是无边无尽的黑夜中偶尔闪过的那层层叠叠、绚丽多彩的极光。极光并不多见，它的每一次出现都会给极地的队员们带来惊喜和欢乐。同时，极光那种极富动感的形象也让我们萌生了不仅让"极光"出现，也让基地队员们自己去模拟极光，通过人与光之间的互动丰富枯燥的极夜生活的想法。

材料构成 Material

设计草图 Sketches

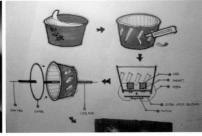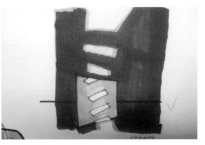

制作过程 Construction

成果表现 Realization

赵无极印象
An impression on Zao Wou-ki

作者 Designers

黄嘉萱　　丁思岑　　吴　林　　殷　明　　蒋竹翌

概念原理 Concept

我们注意到长城站的常见生活垃圾——泡面和乌江榨菜包的内侧镜面纸。将光通过镜面纸反射到纸张表面时，会呈现特殊的光影效果，魔幻而绚丽，其特殊的肌理给我们以启发，想到赵无极先生的山水画，那色彩变幻、笔触有力、富有韵律感和光感的新的绘画空间和我们得到的效果有异曲同工之妙，也能为南极的工作人员的生活带来色彩和乐趣。于是我们尝试用不同的方式将镜面纸揉皱，利用纸面上呈现的不同肌理变化和不同颜色 LED 灯光的排布来塑造出赵无极山水画的感觉。

材料构成 Material

   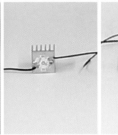 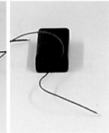

设计草图 Sketches

 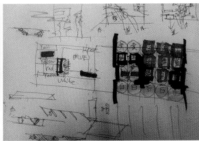 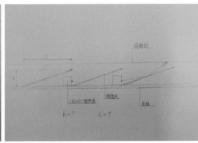  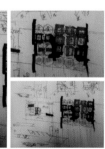

制作过程 Construction

成果表现 Realization

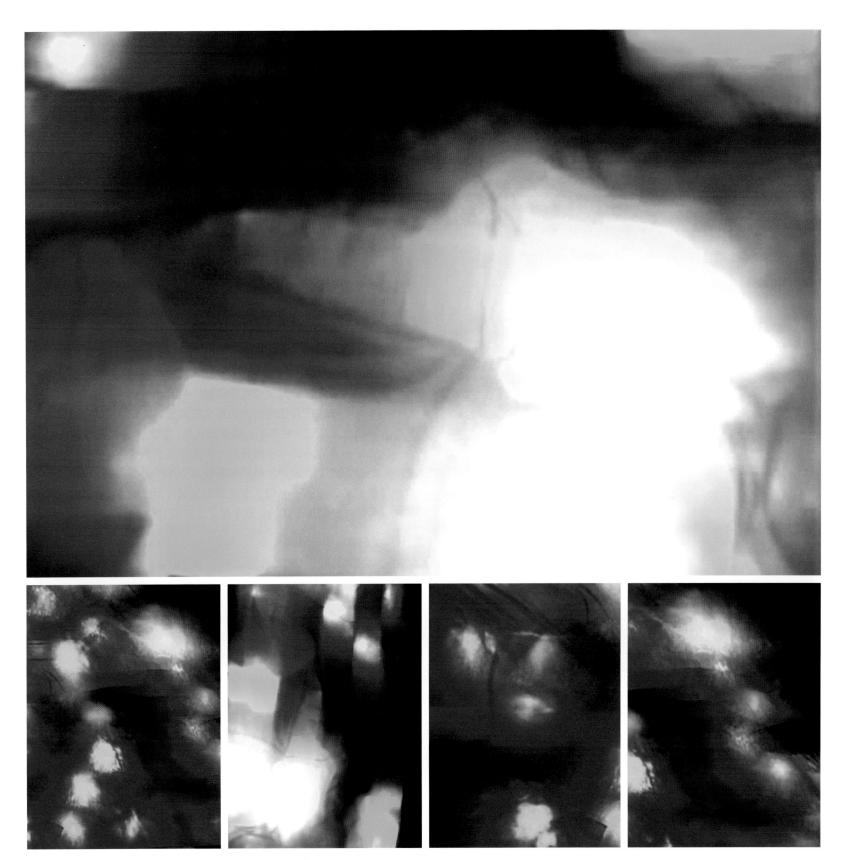

学生感言 Students' comments

学生：李静思 刘含 崔婧 梁芊荟 徐厚哲 薄尧 欧米尔

首先让我感触比较深的就是材料与光的结合的不确定性，生活中常见事物，尤其生活垃圾在与光的结合中也能产生让人想象不到的效果。以光学的视角发现生活中的美应该算是一次不错的体验吧。

还有就是确定一个主题就坚持深化的过程。团队在合作过程中有挫折也有欢乐，有不停的试验也有紧张的展示。整个团队为了一个目标努力，最终达到了自己想追求的效果。

建筑光学的课程作业——制作光艺术装置让我们在欢乐中学习，让我们在学习中欢乐。

学生：沙千陌 常琬悦 罗富缤 刘鹤 叶天娇 郑圭东 傅雷格 承晓宇

光装置既以装置本身的形态，又以发散光为主导，采取最简洁的设计方法。就地取材，南极的易拉罐的回收方式会采取压扁它节约空间的方式，我们不改变人们习惯性的挤压方式，将易拉罐割出几个竖条，利用金属材质本身的弹性和反光特性，扭转一定角度，内外配以彩色光，形成偏心星状图案，组合在一起又似花草效果。在互动方面，我们用铃铛挂在易拉罐中央，在按压其的同时光随着易拉罐上出光口的粗细而发生明暗变化，铃铛也因振动发出声响，增强趣味性，声光变换同在。

我们组之前一直在用纸盘子做设计，但是出来的效果感觉不是光本身的形态，而是照亮了装置，我们组果断舍弃原来的方案，新方案的出现让我们感受到有时候这种类似工业设计的装置，灵感就在一瞬间，实践出真知，有时候空谈是没用的，动手做了才知道效果好不好。

学生：金勋 刘子晨 李方芳 刘茜 路加

这次针对南极极昼下的 LED 灯光装置设计中，我们小组进行了多方面的讨论，最终制作了一个 600mm×600mm×600mm 见方的装置，它四面由 3 层局部镂空的材质组成，一面开口，一面覆盖，可在开口面、覆盖面及装置内部配置各种设定好的光源，在四周投影面上形成多彩斑斓的效果。

在整个制作过程和最后的展示中，我们利用多彩的 LED 灯形成多样的视觉效果，在昼夜的不同情形下也对灯光的视觉感受有了一定的体验与认知，通过装置的完成也思索了不同材质的多样处理对光效的影响，这些都是我们这次作业颇多的内在收获。

学生：马曼·哈山 张辰 马潇潇 薛洁楠 王唯渊 张黎婷 马敏洁 张晗婧

比起给南极科考队员带来光的视觉盛宴，我们组的出发点更看重的是如何让人真正参与到整个光装置中，能丰富他们的视觉更能增加那里的生活情趣。

里面的人在装置里随意自在地活动，外面的人享受着这光影的变化，声音的律动，以及温馨的氛围。人与光能共同勾画出一幅绚丽清新的画面，在极夜享受被太阳包围的暖意。

本次设计，我们班由于人少所以整个班为一个小组，在合作中迸发灵感，在合作中促进友谊，在合作中增进班级凝聚力。

光装置设计——最意外是我们通过努力真的也能做出一个精彩的光艺术作品；最开心是我们表达出了原汁原味的设计理念；最满足是不少观众（评委、其他组、路人）肯定了我们的装置——绚丽的灯光，温馨的表演；最感慨是它必将成为我们大学时代青春岁月里的那一抹光辉，念念不忘。

学生：黄芫 顾倩 姜晗笑 范雅婷 钱静 刘一萱

这次光学作业的出发点非常有实际意义，是为南极科考队员做一个光装置。老师为我们介绍了许多南极的情况，大家也都热情高涨希望能为科考队员的生活带来一些乐趣。

我们在构思、制作，以及展示的过程中都学到了许多平时课本、课堂里学不到的东西——我们认识到了理想与现实的差距，因为我们的构想与实际制作存在很大的偏差；我们也体会到了发现问题到想各种办法解决问题的过程。通过这次光装置的制作，我们不仅学到了如何搭配灯光的色彩，也切身体会到了变废为宝的神奇。更重要的是感受到了团队合作的乐趣。

学生：孙一桐 何凌芳 奚祎婷 潘逸瀚 彭书勉 闵睿

这次光学装置的设计和制作是一次难忘的经历。

刚开始，我们小组打算从南极的特点和科考队员的心理需求出发，试图设计出能够有互动，有变化，能够改善南极科考单调乏味的生活的装置，但是几次讨论都没有什么实质性的进展。于是，我们把南极科考队员的生活用品都买来，用灯光试一试。在尝试的过程中，我们突然发现，方便面包装袋背面反射光投在墙面的效果特别绚丽，有银河般的奇妙效果。接着通过小组的讨论深化，我们增加了互动，设计了几种配色的方案，使得原有的效果更加丰富。

这次装置的设计，是与我们平时设计的思路完全不同的一次"设计"。它是从材料出发，从细部出发，然后挖掘材料与光的表现力。在这种亲自动手尝试的过程中，我们加深了对于光的理解，感受到了光的丰富的表现力，懂得了团队合作的重要性，学习到了许多课本以外的知识。

学生：曹诗敏 韩雪松 黄艺杰 胡佳林 夏孔深 宋睿

很多人都说23组运气好，随便捡个瓶子效果竟然还不错，真正制作只用了两天，还没有熬夜。

其实从任务布置下来到吃饭、走路、做模型的时候我们都在构思，第一次头脑风暴确定了概念：不需送任何装置，只带想法过去；装置除视觉效果外还要实用、互动、环保；将春天带到被寒冬包围的南极。

第二次风暴，我们在4个草模中确定了第一个方案：用丝袜相互缝合成中间有孔洞的奶酪状（类似伊东丰雄的台中大都会歌剧院概念模型），人拉扯时不同颜色的光会相互交融。

第一次展示的前一晚，我们在南门外便利店开始做模型，做到12点，我们开始产生怀疑：我们要做的是通过一个装置展现光的独特，而不是用光来表现一个模型，经过一个多小时的讨论后，我们毙掉了这个方案。没了方案后，我们围坐在桌子旁焦急地苦想，气氛很紧张，好像一点燃就会爆炸。

后来我们开始在超市里找材料，顾不得超市收银员怪异的目光。组员韩雪松从冰柜里拿出两瓶饮料，用LED灯照时发现效果很好，像花朵绽放一样。我们又激动地到冰柜里拿了十多种饮料试验。最后顺利地定下了灯光装置的展示与使用方式。

Party前后我们并不顺利：展位被汽车挡住，精心制作的LED帘子放在专教忘了拿下来……但是我们一直情绪高涨，为自己能和南极科考队员联系起来而骄傲。同时我们也收获很多：组员韩雪松进入实验班后，常常感到自卑，但他从这次作业中找到了自信；视频中两个男生搭架子时互相擦汗，女生在男卫生间洗瓶子，大家可能认为是搞笑，但这是时间紧、迫不得已时的真实画面，新组成的班级顿时融洽了许多。

不论是飘絮洒落，还是繁花盛开，我们希望能通过自己的努力让长城站花团锦簇，让科考队员心花怒放。

现场评审 Site assessment

评委老师：章明

郝洛西教授指导的光艺术装置实验，每次作业汇报都像一场盛大的节日欢宴，学生的作业在黑暗中盛开在建筑学院的各个楼宇间，给人留下极其深刻的印象。

从建筑学教育的角度来说，"光艺术装置"实验极大地挖掘了学生们的创造力和想象力，这也是建筑学学生需要培养、修炼的本领。这个实验完全突破了传统的刻板教学，通过这一实验训练，学生们不仅在创造力、实践能力方面得到极大提升，更重要的是，学生们更深刻地认识到现代建筑要素的复杂多元性，灯光在建筑中的作用已不是简单的点缀提升作用，更能成为建筑创作的灵感来源。

我已参加了"光艺术装置"实验五届的正式汇报，在感叹学生们的丰富创造力的同时，也特别佩服郝洛西教授的教学团队对此付出的心血。我们的建筑学学生每届都是百余人，每次实验操作也不是一两个学时就能解决的。郝教授的教学团队在实验进行的两个星期内，几乎每天晚上都要对学生进行辅导。此外，他们还积极与社会单位、企业沟通联系，多个学生作业已在世博会等重大场合得以应用，据悉2013年的成果还将运往我国南极科考站。正是教师们的无私付出、学生的全情投入才取得实验如此丰硕的成果。希望该实验能得到更大的支持，在后续的教学中取得更佳的教学成果。

评委老师：刘颂

郝洛西教授多年来投入大量心血不断创新、完善光艺术装置实验课程。实验课程突破了传统物理教学的模式，学生在制作实验作品的过程中自然地完成对知识的学习和探究。整个课程安排极大调动了学生的主观能动性，真正做到了寓教于乐。实验内容紧跟科技前沿，要求学生的实验作品必须与 LED 光源特性高度结合，每个实验小组都必须在规定时间内，围绕实验目的完成"立意""寻材""设计""制作""演示"等一系列工作，极大地培养了学生之间的团队协作能力。通过本实验，学生们能够深入掌握相关的光学知识及 LED 光源工作原理，初步了解照明控制系统的设计思路，极大地培养了学生对项目的整体控制能力。

实验成果展示部分充分调动了学生们的积极性、创造性，每组作品创意独特、效果绚丽，成为我院每年一度的光艺术盛宴。多年来赢得老师、同学们的一致好评。

本实验课程教学手法新颖、内容深刻却不乏趣味性，深受广大学生喜爱。

评委老师：阴佳

郝洛西教授的"光艺术装置教学实验"突破了传统的单纯物理光学实验课程模式，与建筑的设计与材料紧密结合，使学生在实验过程中不仅认知材料与光和建筑形体与光影的互为影响及作用，更重要的是使学生不再将"光"仅仅作为物理想象而是建筑设计中至关重要的形式要素。如何学会"发现"，如何在设计中审美挖掘和主动驾驭"光"，如何打破传统的思维模式制约去利用材料，是这一教学实验环节的核心。这种在实验教学中嵌入艺术"基因"和强调"发现"眼光的创新实践必将会对学生知识结构补充和完善产生久远的作用，同时，也开启了一扇新的认识世界的窗口。

这项实验的教学已有十年的探索和完善历程，郝教授与她的教学团队始终在研究与调整教学模式并将照明领域最新的科学技术融入其中。从最传统的光照控制模式，到音控与光艺术结合，再到 LED 绿色照明技术及媒体化艺术创作，使得课程始终与社会发展同步，并将节能减耗作为创作理念，让学生去发现和寻找生活中的废弃物品，使这些弃旧物与光结合之后绽放出耀眼的光彩。

这门以创新思维为指导前提的实验课程，不仅已成为学院的品牌课程，更使得我们学校在这一领域中的教学处于中国高校的前沿。

143

历年教学回顾 Review of years of teaching

| 2002 | 2003-2005 | 2006-2007 | 2008 |

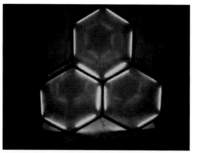

光与水立方——北京奥运会国家游泳中心（水立方）室内及立面光环境设计

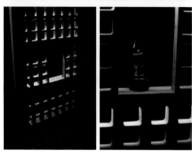

课程要求学生对生活中的光环境进行调研，利用实验室的设备，采用不同的材料和照明方式，截取有意思的片段进行表现。这是本课程体系首次尝试脱离常规的教学模式，让学生亲自体验光照场景，理解物理量的视觉意义，以培养他们运用所学知识和设计技法解决实际问题的能力。

西安汉阳陵遗址保护展示厅及周边环境照明设计

在建筑光环境实验性教学体系探索初期，光艺术装置教学是照明专门化毕业设计中的重要环节。学生们根据自己的方案制作整个建筑或局部片段的缩尺灯光工作模型，在建立光环境概念的同时，掌握第一手光度数据资料，更好地将光度物理量与实际效果相联系。

利用传统光源如白炽灯、卤素灯、荧光灯等，易于市场选购的光源和实验室的设备，将光作为生动的设计语汇进行创作。其形态、光影、明暗构成动态或静态的光照图示。学生最终将成果制作成幻灯片或多媒体动画，展示他们对光与材质、空间的认识和理解。

由于照明技术的发展，光艺术装置实验环节引入低碳环保的LED光源作为创作素材。教学场地的LED显示屏成为同学们的创作基础，每组学生分配25cm×25cm大小的显示屏，他们在充分了解LED发光特点、材料特性的基础上进行光艺术媒体界面的创作。

2009　　　　　　　2010　　　　　　　2011–2012　　　　　　2013–2015

2010世博会前期的学生作业，在世博文化中心两个入口大厅背景墙得到深化应用，最终作品完全由师生现场制作完成，获得了应用单位的高度认可，取得了良好的社会反响。并在若干年后被应用于实验室人居空间健康照明的研究课题中。

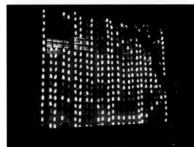

"声光SHOW"大型装置中，学生们在了解语音输入、声音信号识别、人机交互等原理的基础上，自己编排声光效果，再利用矿泉水桶作为发光像素点，上演了一场别开生面的"声光秀"。

课程更加注重实践性，学生亲自动手焊接LED颗粒并连接电源。装置作业的规模、成熟度及艺术效果得到了加强。并在最后加入评审展示的环节，评审老师根据展示现场效果给出课程成绩。

从2013年开始装置作业有了主题，学生们在限定主题、场地、时间、装置制作材料、LED颗粒颜色和数量的条件下，以小组为单位，制作光艺术装置，结合"徒手控制"的动态编排，探索光影如何与空间、环境、音乐互动，探讨光与色彩如何改善生活环境、诠释传统文化、演绎戏剧情节。发现光的多维度应用，体验光的无限魅力。

图书在版编目（CIP）数据

2013·南极之光 / 郝洛西等著 . -- 上海：同济大学出版社，2016.6
（同济大学建筑学专业"建筑物理（光环境）"教学成果专辑）
ISBN 978-7-5608-6200-2

Ⅰ.①2… Ⅱ.①郝… Ⅲ.①建筑物理学—教学研究—高等学校—文集
②建筑光学—教学研究—高等学校—文集 Ⅳ.①TU11-53

中国版本图书馆 CIP 数据核字 (2016) 第 025194 号

同济大学建筑学专业"建筑物理（光环境）"教学成果专辑
2013·南极之光

郝洛西 等 著

责任编辑	张　睿
责任校对	张德胜
装帧设计	李　丽
APP 制作	今尚数字
出版发行	同济大学出版社（www.tongjipress.com.cn）
地　　址	上海四平路 1239 号（200092）
电　　话	021-65985622
经　　销	全国各地新华书店
印　　刷	上海丽佳制版印刷有限公司
开　　本	787mm×1092mm　1/12
印　　张	37.33
字　　数	940000
版　　次	2016 年 6 月第 1 版
印　　次	2016 年 6 月第 1 次印刷
书　　号	ISBN 978-7-5608-6200-2
定　　价	210.00 元（全三册）